GCE A-Level

Pure and Applied Mathematics

R. A. Parsons, B.Sc.

and

A. G. Dawson, B.Sc.

Published by Charles Letts Books Ltd
London, Edinburgh and New York

Published 1983 by Charles Letts Books Ltd
Diary House, Borough Road, London SE1 1DW

1st edition 1st impression

Made and printed by Charles Letts (Scotland) Ltd
ISBN 0 85097 559 X

TONY DAWSON graduated from Southampton University with an honours degree in Mathematics. After completing his post-graduate Certificate in Education, also at Southampton University, he began teaching mathematics at Wallington Grammar School for Boys. Since then he has taught the subject to all levels of ability including Oxbridge Entrance and his teaching has covered both modern and traditional syllabuses. He has had considerable experience as a Head of Department at Woking Grammar School for Boys. More recently he has been appointed Deputy Principal of Woking Sixth Form College but is still actively involved in preparing candidates for advanced level examinations.

ROD PARSONS graduated from Bristol University and completed a post-graduate certificate in Education at Southampton University. Since then he has taught in schools and sixth form colleges in Surrey. He has taught both modern and traditional syllabuses, has written for S.M.P. and lectured to teachers at S.M.P. conferences. He was Head of the Mathematics Department at Sunbury Sixth Form College and after that Head of Mathematics at Woking Sixth Form College. He is now an Assistant Principal at Esher Sixth Form College.

Worked Examples

APPLIED MATHEMATICS, E. M. Peet, B.Sc.

BIOLOGY, A. G. Toole, B.Sc.

CHEMISTRY, J. E. Chandler, B.Sc.
and R. J. Wilkinson, Ph.D.

ECONOMICS, R. Maile, B.A.

GEOGRAPHY, C. Lines, M.Sc.

PHYSICS, G. M. George, B.Sc.

PURE MATHEMATICS, R. A. Parsons, B.Sc.
and A. G. Dawson, B.Sc.

PURE AND APPLIED MATHEMATICS,
R. A. Parsons, B.Sc. and A. G. Dawson, B.Sc.

Contents

Introduction

This book of worked examples is designed as a sequel to *Pure and Applied Mathematics* by the same authors in the Key Facts series and is set in the style of advanced level mathematics examination questions.

The chapters follow the same headings as those in the Pure and Applied Mathematics book. Most examination syllabuses have a common core of content but may vary considerably in the range and depth of the extra topics which they include. The book attempts to illustrate examples of different styles of examination question as appropriate in the various chapters and these have been grouped into the following categories:

Multiple choice questions (Type A) Select the correct answer
In this type of question five possible responses are given and the candidate is required to decide which one is correct.

Multiple choice questions (Type B) Answer according to the table

A	**B**	**C**	**D**	**E**
1, 2, 3 correct	1, 3 only	2, 3 only	2 only	3 only

Three statements are given, each of which may be true or false. Each statement should be considered separately and the final choice made according to the above table.

General questions Some Boards set two or three papers and these may consist of a section A containing 8 to 12 short questions followed by a choice of longer questions or alternatively a whole paper may involve shorter type problems. These usually test an application of just one topic or idea.

Examples of **longer questions** are included to illustrate the type of question the candidate has to answer when faced with a choice of, say, four out of six or seven out of eleven. These questions generally contain two or three applications on one theme or may require processes from different branches of the subject.

The examples we have presented are designed to cover most of the topics examined on an A-Level syllabus for Pure and Applied Mathematics. These are based on our experience of teaching Advanced Level Mathematics over many years while preparing students for different examination boards.

As teachers, we recognize that we are leading students to understand and appreciate mathematics more fully and not just preparing them for examinations. Mathematics is an art and possesses a beauty of its own which gives satisfaction to those who master it. However, it is also a tool required in many other areas and subjects, and teachers will attempt to adequately cover both of these aspects.

Mathematics is, by its nature, a logical subject and we would encourage students to set out their answers in a clear and coherent manner. These examples will give some idea of the way in which solutions should be presented.

Although we feel that coaching specifically for examinations is not educationally very sound there can be no doubt that 'practice makes perfect' and a worked example may be a distinct aid to learning.

We would advise students to use this book constructively and would suggest that they try to answer the questions before referring to our explanations. Sometimes, however, some guidance may be necessary from the outset and, if this is the case, the student should read the first few lines of the solution and then try to carry the rest of the argument through for himself.

It may well be helpful to cover the working with a card and only reveal each successive line of explanation after trying that stage for oneself.

Mathematics is a subject that cannot easily be mastered by reading alone. It is vital for students to have paper and pencil available so that the steps of the working can actually be written out. Note that, while we have endeavoured to include all the necessary working, students may need to amplify certain stages to fully understand the solutions.

It is good practice to make full use of sketches and diagrams when writing out solutions. For reasons of space we have kept the number of diagrams to a minimum but we would encourage students to draw them for themselves.

Finally, since the style of examination papers and questions varies from Board to Board we would suggest that students spend some time during their examination preparation familiarizing themselves with some of the past examination papers set by their own particular examination board.

Chapter 1
Algebra

Multiple choice questions (Type A) Select the correct answer

Example 1
When $f(x) = x^3+2x^2+3x+4$ is divided by $x+2$ the remainder is

A 2 **B** -2 **C** 14 **D** 24 **E** 0

This is a straightforward question which is best answered by using the remainder theorem. The remainder is given by $f(-2)$.

$$\text{Remainder} = f(-2) = (-2)^3+2(-2)^2+3(-2)+4$$
$$= -8+8-6+4 = -2$$
ANSWER B

Example 2
The sum of the roots of $2x^3-3x^2+4x-5=0$ is equal to

A -3 **B** 3 **C** $-1\frac{1}{2}$ **D** $1\frac{1}{2}$ **E** $-2\frac{1}{2}$

If the roots (answers) of the equation $ax^3+bx^2+cx+d=0$ are α, β and γ, then

$$\alpha+\beta+\gamma = -\frac{b}{a}; \quad \alpha\beta+\beta\gamma+\alpha\gamma = \frac{c}{a}; \quad \alpha\beta\gamma = -\frac{d}{a}$$

The sum of the roots of the equation is equal to $-\dfrac{(-3)}{2} = 1\frac{1}{2}$.
There is no need to solve the equation completely. ANSWER D

Example 3
If $\dfrac{5}{x^2-x-6} \equiv \dfrac{a}{x+2}+\dfrac{b}{x-3}$ then

A $a=1$ **B** $b=1$ **C** $a=2$ **D** $b=3$ **E** $a=-2$

The aim of this question is to express $\dfrac{5}{x^2-x-6} = \dfrac{5}{(x+2)(x-3)}$ in partial fractions.

Assuming $\dfrac{5}{(x+2)(x-3)} = \dfrac{a}{x+2} + \dfrac{b}{x-3}$ and multiplying both sides by $(x+2)(x-3)$ gives $5 = a(x-3)+b(x+2)$.

Let $x=3 \Rightarrow 5=5b \Rightarrow b=1$; Let $x=-2 \Rightarrow 5=-5a \Rightarrow a=-1$.

Alternatively, the value of a can be found by substituting $x=-2$ in $\dfrac{5}{(x+2)(x-3)}$ and ignoring the bracket $(x+2)$ whose value is zero.

The value of b is found by putting $x=3$ and ignoring $(x-3)$. Therefore, $a = \frac{5}{-5} = -1$ and $b = \frac{5}{5} = 1$.

ANSWER B

Multiple choice questions (Type B) Answer according to the table

A	**B**	**C**	**D**	**E**
1, 2, 3 correct	1, 3 only	2, 3 only	2 only	3 only

> **Example 4**
> A curve is given in parametric form as $x=2-3t, y=t+4$.
>
> **1** It is a straight line **2** Its gradient is $\frac{1}{3}$
> **3** Its intercept on the y-axis is $4\frac{2}{3}$

The best way to answer statement 1 is to try to eliminate t and form the Cartesian (x, y) equation of the curve.

$x+3y=2-3t+3t+12=14 \Rightarrow x+3y=14$ (a straight line)

Rearranging the equation gives $y=-\frac{1}{3}x+4\frac{2}{3}$ which represents a straight line with gradient $-\frac{1}{3}$ and intercept on the y-axis $4\frac{2}{3}$.

Statements 1 and 3 are true giving answer B. ANSWER B

> **Example 5**
> For the expression x^3+y^3
>
> **1** $x+y$ is a factor **2** $(x+y)^3$ is equivalent
> **3** x^2-xy+y^2 is a factor

$x^3+y^3=(x+y)(x^2-xy+y^2)$ so statements 1 and 3 are true.
$(x+y)^3=x^3+3x^2y+3xy^2+y^3$ so statement 2 is false.

ANSWER B

Example 6
The region satisfied by the three inequalities contains one point only with integer coordinates.

1 $y > x^2 - 2x,\ x + y < 2,\ y > x$ **2** $y > x^2 - 2x,\ x + y < 2,\ y < x$

3 $y > x^2 - 2x,\ x + y > 2,\ y < x$

The diagram in Figure 1 shows the graphs of the three equalities

$y = x^2 - 2x$

$x + y = 2$

$y = x.$

Statement 1 is represented by region Q.

Statement 2 is represented by region P.

Statement 3 is represented by region R.

All three regions contain one integer point only.

ANSWER A

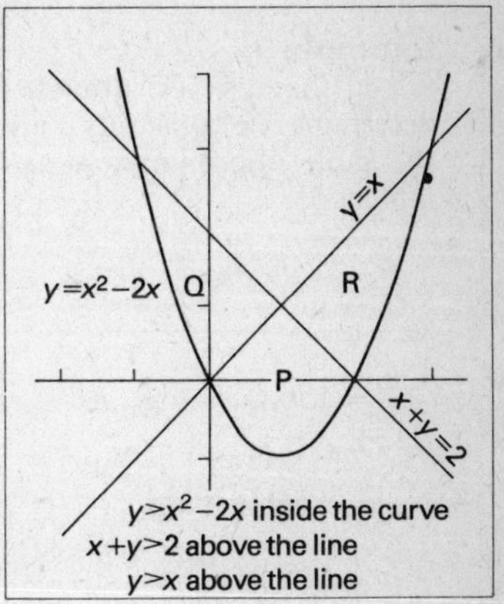

$y = x^2 - 2x$ Q R P $y = x$ $x + y = 2$

$y > x^2 - 2x$ inside the curve
$x + y > 2$ above the line
$y > x$ above the line

Figure 1

General questions

Example 7
Find the set of real values of x for which $x^2 - 9x + 20$ is negative.

$x^2 - 9x + 20 = (x - 4)(x - 5)$ which takes zero values when $x = 4$ and $x = 5$. The graph of $x^2 - 9x + 20$ is a positive quadratic; the positive x^2 term implies a U-shaped curve with a minimum point. The curve cuts the x-axis at (4,0) and (5,0) and lies below the x-axis for values between these points.
The solution set is $4 < x < 5$ ($x = 4$ and $x = 5$ are excluded).

Example 8
Find the set of values of k for which $x^2 + kx + 9$ is positive for all real values of x.

The graph of x^2+kx+9 is a positive quadratic (U-shaped) which will be positive for all real values of x if it lies completely above the x-axis. This will happen if there are no real roots to the equation $x^2+kx+9=0$.

Using the quadratic formula $x=\dfrac{-b\pm\sqrt{b^2-4ac}}{2a}$ no real roots occur when $b^2<4ac \Rightarrow k^2<36 \Rightarrow -6<k<6$

Example 9
Given that $y=ax^n$ draw a suitable straight line graph to determine the constants a and n assuming that the values in the table approximately satisfy the relation (a and n are integers).

x	1	2	3	4	5
y	1.9	16.2	53	130	240

$y=ax^n \Rightarrow \log_{10}y=\log_{10}ax^n$

$\Rightarrow \log_{10}y=\log_{10}a+n\log_{10}x$

Plot $\log_{10}x$ against $\log_{10}y$

$\log_{10}x$	0	0.3	0.48	0.6	0.7
$\log_{10}y$	0.28	1.21	1.72	2.11	2.38

These values are plotted in Figure 2.
The gradient (n) is

$$\frac{2.1}{0.7}=3=n$$

The intercept on the y-axis ($\log_{10}a$) is $0.29 \Rightarrow a=1.95$.
Since a is an integer, $a=2$.

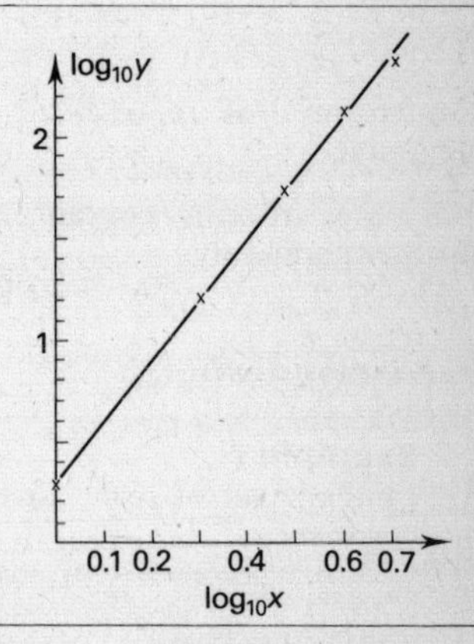

Figure 2

Example 10
If p and q are the roots of $2x^2+3x+4=0$ find

(a) $\dfrac{1}{p}+\dfrac{1}{q}$ (b) p^2+q^2 (c) p^3+q^3

(a) It is not necessary to find the exact roots (in fact the roots are not even real).

If p and q are the roots of $ax^2+bx+c=0$

then $p+q = -\frac{b}{a} = -1\frac{1}{2}$ and $pq = \frac{c}{a} = \frac{4}{2} = 2$

$$\frac{1}{p}+\frac{1}{q} = \frac{p+q}{pq} = \frac{-1\frac{1}{2}}{2} = -\frac{3}{4}$$

(b) $p^2+q^2 = (p+q)^2-2pq = (-1\frac{1}{2})^2-4 = 2\frac{1}{4}-4 = -1\frac{3}{4}$

(c) $p^3+q^3 = (p+q)^3-3p^2q-3pq^2$

$$= (p+q)^3-3pq(p+q)$$

$$= (-1\tfrac{1}{2})^3-6(-1\tfrac{1}{2})$$

$$= -\tfrac{27}{8}+9 = 5\tfrac{5}{8}$$

These questions are usually exercises in algebraic manipulation. They are not difficult but require practice in expressing the quantities required in terms of $p+q$ and pq.

Example 11

Given that $\log_2 x+2\log_4 y = 4$, show that $xy = 16$. Hence solve for x and y the simultaneous equations

$$\log_{10}(x+y)=1 \qquad \log_2 x+2\log_4 y=4$$

$\log_4 y = \frac{\log_2 y}{\log_2 4} = \frac{1}{2}\log_2 y$ using change of logarithm base.

$$\log_2 x+2\log_4 y = 4 \Rightarrow \log_2 x+\log_2 y = 4 \Rightarrow \log_2 xy = 4$$

This is equivalent to $xy = 2^4 = 16$ (1)

The first equation is equivalent to $x+y = 10^1 = 10$ (2)

Substituting for y from (1) into (2) $\quad x+\frac{16}{x} = 10$

$\Rightarrow x^2+16 = 10x \Rightarrow x^2-10x+16 = 0 \Rightarrow x = 2$ or $x = 8$ by factors.
$x = 2, y = 8$ or $x = 8, y = 2$ are the two pairs of solutions.

The equations (1) and (2) are symmetrical in x and y leading to the symmetrical solutions.

This type of question requires knowledge of the laws of logarithms:

Basic definition $\quad \log_a b = c \Leftrightarrow a^c = b$ or $2^3 = 8 \Leftrightarrow 3 = \log_2 8$

Change of base $\quad \log_a b = \frac{\log_c b}{\log_c a}$, also $\log_a x+\log_a y = \log_a xy$

Example 12

Express in partial fractions $\dfrac{2}{(x+1)^2(x^2+1)}$

$$\frac{2}{(x+1)^2(x^2+1)} = \frac{ax+b}{x^2+1} + \frac{c}{x+1} + \frac{d}{(x+1)^2}$$

These include all the possible factors. a, b, c and d could be zero. Multiplying both sides by $(x+1)^2(x^2+1)$ gives

$$2 = (ax+b)(x+1)^2 + c(x+1)(x^2+1) + d(x^2+1)$$

$x = -1 \Rightarrow 2 = 2d \Rightarrow d = 1$

Equating coefficients of $x^3 \Rightarrow 0 = a+c$ (1)
Equating coefficients of $x^2 \Rightarrow 0 = 2a+b+c+d$ (2)
Equating coefficients of $x \Rightarrow 0 = a+2b+c$ (3)
Equating constant terms $\Rightarrow 2 = b+c+d$ (4)

(3) − (1) $\Rightarrow b = 0$ and from (4) $2 = 0+c+1 \Rightarrow c = 1 \Rightarrow a = -1$

So $\dfrac{2}{(x+1)^2(x^2+1)} = \dfrac{-x}{x^2+1} + \dfrac{1}{x+1} + \dfrac{1}{(x+1)^2}$

Note that the first statement is crucial; a denominator of x^2+1 requires a numerator of $ax+b$ and both parts of a repeated factor are needed.

Example 13

Solve the equation $3^x \times 2^{2x} = 7$

Either, taking logs to base 10 (although any base will do)

$$\begin{aligned} x\log 3 + 2x\log 2 &= \log 7 \\ x(\log 3 + 2\log 2) &= \log 7 \\ x(\log 3 + \log 4) &= \log 7 \\ x\log 12 &= \log 7 \\ x &= \frac{\log 7}{\log 12} = \frac{0.845}{1.079} = 0.783 \end{aligned}$$

or $2^{2x} = 4^x \Rightarrow 3^x \times 4^x = 12^x = 7$

$$\Rightarrow x\log 12 = \log 7 \Rightarrow x = \frac{\log 7}{\log 12} = 0.783$$

Example 14
Given that $\log_{10}2 = 0.301$ and $\log_{10}3 = 0.477$ find the values of $\log_{10}4$, $\log_{10}5$, $\log_{10}6$, $\log_{10}2\frac{2}{3}$ and $\log_{10}2\frac{7}{9}$.

$$\log_{10}4 = \log_{10}2^2 = 2\log_{10}2 = 2\times 0.301 = 0.602$$
$$\log_{10}5 = \log_{10}\tfrac{10}{2} = \log_{10}10-\log_{10}2 = 1-0.301 = 0.699$$
$$\log_{10}6 = \log_{10}2+\log_{10}3 = 0.301+0.477 = 0.778$$
$$\log_{10}2\tfrac{2}{3} = \log_{10}\tfrac{8}{3} = \log_{10}8-\log_{10}3 = 3\log_{10}2-\log_{10}3$$
$$= 0.903-0.477 = 0.426$$
$$\log_{10}2\tfrac{7}{9} = \log_{10}\tfrac{25}{9} = \log_{10}25-\log_{10}9 = 2\log_{10}5-2\log_{10}3$$
$$= 1.398-0.954 = 0.444$$

Example 15
If $g(x) = x^3-2x^2-5x+6$

(a) solve $g(x)=0$ (b) sketch the graph of $g(x)$
(c) find any maximum, minimum or inflexion points

(a) Possible factors of $g(x)$ are $(x\pm 1)$, $(x\pm 2)$, $(x\pm 3)$, $(x\pm 6)$.
Use the factor (remainder) theorem to test for factors.
$g(1) = 1-2-5+6 = 0 \Rightarrow (x-1)$ is a factor.
By inspection (or division) $g(x) = (x-1)(x^2-x-6)$
$= (x-1)(x+2)(x-3)$
$g(x) = 0 \Rightarrow x = 1$ or $x = -2$ or $x = 3$.

Note that $g(3) = 27-18-15+6 = 0 \Rightarrow (x-3)$ is a factor.
$g(x) = (x-3)(x^2+x-2)$
$= (x-3)(x-1)(x+2)$
and the factors are unique no matter in which order you find them.

(b) $g(x)$ is a positive cubic (positive x^3 term $\Rightarrow$ graph goes from bottom left hand corner to top right hand corner).

(i) Look for zeros
$g(x)=0$ when $x = 1, -2, 3$.

(ii) $g(0) = 6$ (constant term)

(iii) $g(-1) = -1-2+5+6 = 8$
$g(2) = 8-8-10+6 = -4$

(iv) Do not assume maximum is at $x = -1$ or minimum is at $x = 2$ (see part (c)).

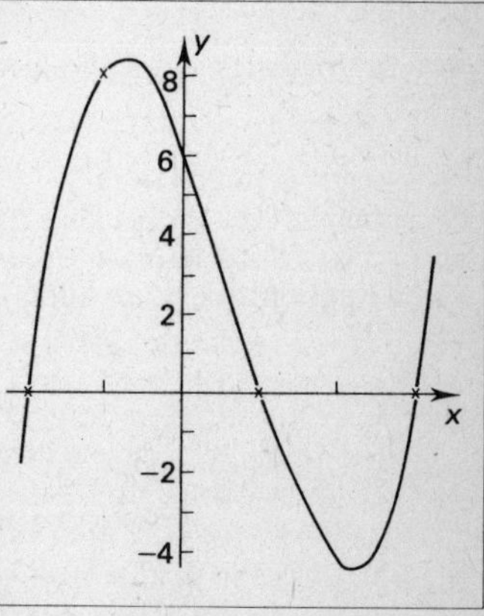

Figure 3

(c) $g'(x) = \dfrac{dg}{dx} = 3x^2 - 4x - 5 = 0 \Rightarrow x = \dfrac{4 \pm \sqrt{16+60}}{6}$

$= 2.12$ or -0.79.

Knowing $g(x)$ is a positive cubic gives $x = 2.12$ as a minimum point and $x = -0.79$ as a maximum point.

Check with $g''(x) = \dfrac{d^2g}{dx^2} = 6x - 4$

$g''(2.12) = 6 \times 2.12 - 4 = 8.72 > 0 \Rightarrow x = 2.12$ is a minimum point.
$g''(-0.79) = 6 \times (-0.79) - 4 = -0.84 < 0 \Rightarrow$ a maximum point.
Points of inflexion are given by $g''(x) = 0 \Rightarrow 6x - 4 = 0 \Rightarrow x = \frac{2}{3}$.
Note that $g'(\frac{2}{3}) = 1\frac{1}{3} - 2\frac{2}{3} - 5 = -6\frac{1}{3}$ and is not zero.
Note also that in general not all points where $g''(x) = 0$ are points of inflexion.

Example 16
Show that $(x-2)$ is a factor of $x^4 + x^3 - 7x^2 - x + 6$ and solve the equation $x^4 + x^3 - 7x^2 - x + 6 = 0$.

To show that $x - 2$ is a factor, find the value of the expression when $x = 2$. If this is zero, $x - 2$ is a factor.

$2^4 + 2^3 - 7 \times 2^2 - 2 + 6 = 16 + 8 - 28 - 2 + 6 = 0 \Rightarrow x - 2$ is a factor.

This could also be shown by dividing the expression by $x - 2$.

By division or inspection

$f(x) = x^4 + x^3 - 7x^2 - x + 6 = (x-2)(x^3 + 3x^2 - x - 3)$

Now factorize the cubic bracket; $x - 1$ is clearly a factor.

$f(x) = (x-2)(x-1)(x^2 + 4x + 3) = (x-2)(x-1)(x+1)(x+3)$
$f(x) = 0 \Rightarrow x = 2$ or $x = 1$ or $x = -1$ or $x = -3$.

This question is easy because the expression factorized. If the cubic factor does not factorize other methods would have to be used which might prove more difficult.

Chapter 2
Trigonometry

Multiple choice questions (Type A) Select the correct answer

Example 1
$\sin(A+B)-\sin(A-B) =$

A $2\sin A\cos B$ **B** $2\sin A\sin B$ **C** $2\cos A\sin B$
D $2\cos A\cos B$ **E** $-2\cos A\sin B$

Method 1 $\sin(A+B) = \sin A\cos B+\cos A\sin B$
$\sin(A-B) = \sin A\cos B-\cos A\sin B$

Subtracting these two equations gives $2\cos A\sin B$. ANSWER C

Method 2 Using $\sin P-\sin Q = 2\cos\frac{1}{2}(P+Q)\sin\frac{1}{2}(P-Q)$
$P+Q = (A+B)+(A-B) = 2A$ and $P-Q = (A+B)-(A-B) = 2B$

So $\sin(A+B)-\sin(A-B) = 2\cos A\sin B$. ANSWER C

In this question there is little to choose between the two methods. However, in some trigonometrical questions, using the right formula at the right time can save much time and working. It is advisable to learn these formulae thoroughly.

Example 2
$\cos 120° =$

A $\frac{\sqrt{3}}{2}$ **B** $\frac{1}{2}$ **C** $-\frac{\sqrt{3}}{2}$ **D** $-\frac{1}{2}$ **E** $\frac{1}{\sqrt{2}}$

$\cos 120° = -\cos(180°-120°) = -\cos 60° = -\frac{1}{2}$. ANSWER D

Remember to take the angle away from 180° and that the cosine is negative in the second quadrant.

Example 3

If x is acute $\tan x = \frac{a}{b} \Rightarrow \cos x =$

A $\frac{b}{a}$ **B** $\frac{a}{\sqrt{a^2+b^2}}$ **C** $\frac{b}{\sqrt{a^2+b^2}}$ **D** $\frac{a}{\sqrt{a^2-b^2}}$ **E** $\frac{b}{\sqrt{a^2-b^2}}$

Draw a triangle (right-angled) with x as an acute angle, a as the opposite side and b as the adjacent side. Then the hypotenuse is $\sqrt{a^2+b^2}$ and $\cos x = b/\sqrt{a^2+b^2}$. ANSWER C

Example 4
If $f(x) = 3\cos x + 4\sin x$ then the maximum value of $f(x)$ is

A 3 **B** 4 **C** 5 **D** 6 **E** 7

Method 1 $f'(x) = \dfrac{df}{dx} = -3\sin x + 4\cos x = 0$ for maximum

$\Rightarrow 3\sin x = 4\cos x$

$\Rightarrow \tan x = \frac{4}{3}$ (since $\sin x \div \cos x = \tan x$)

$\Rightarrow \sin x = \frac{4}{5}$ and $\cos x = \frac{3}{5}$. There is no need to find x.

$f(\max) = \frac{9}{5} + \frac{16}{5} = \frac{25}{5} = 5.$ ANSWER C

Method 2 $f(x) = 5(\frac{3}{5}\cos x + \frac{4}{5}\sin x)$
Put $\sin y = \frac{3}{5}$ and $\cos y = \frac{4}{5}$,

$$f(x) = 5(\sin y\cos x + \cos y\sin x) = 5\sin(y+x)$$

This will have a maximum value of 5 since the maximum value of $\sin(y+x)$ is 1.

Method 1 is the obvious way to find the maximum value of a function but Method 2 has an important application when solving equations (see Example 24).

Example 5
$\dfrac{d}{dx}(\ln\cos x) =$

A $\dfrac{1}{\cos x}$ **B** $\dfrac{\sin x}{\cos x}$ **C** $-\dfrac{1}{\sin x}$ **D** $-\tan x$ **E** $\sec x$

This is a composite function (or function of a function). The derivative of $\ln y$ is $\dfrac{1}{y}$ and in this example the logarithm is differentiated first to give $\dfrac{1}{\cos x}$ which is then multiplied by the derivative of $\cos x$, namely $-\sin x$.

The result is $-\dfrac{\sin x}{\cos x} = -\tan x.$ ANSWER D

Some students forget the way to differentiate a composite function and produce answer A (answer E is the same as answer A) or confuse the two processes (differentiating the log and the cos) to produce answer C.

Example 6
$\tan(60° + 45°) =$

A $\dfrac{\sqrt{3}+1}{1-\sqrt{3}}$ **B** $\dfrac{1+\sqrt{3}}{\sqrt{3}-1}$ **C** $\dfrac{\sqrt{3}-1}{\sqrt{3}+1}$ **D** $\dfrac{1-\sqrt{3}}{1+\sqrt{3}}$ **E** none of these

$$\tan(60° + 45°) = \frac{\tan 60° + \tan 45°}{1 - \tan 60° \tan 45°} = \frac{\sqrt{3}+1}{1-\sqrt{3}\times 1}$$ ANSWER A

It is worthwhile learning the ratios associated with the 30°,60°,90° and 45°,45°,90° triangles. Examiners often ask questions on these triangles.

$\sin 30° = \cos 60° = \frac{1}{2}$; $\sin 60° = \cos 30° = \dfrac{\sqrt{3}}{2}$; $\tan 30° = \dfrac{1}{\sqrt{3}}$;
$\tan 60° = \sqrt{3}$

Multiple choice questions (Type B) Answer according to the table

A	**B**	**C**	**D**	**E**
1, 2, 3 correct	1, 3 only	2, 3 only	2 only	3 only

Example 7
Given that $\tan A = t$

1 $\sin 2A = \dfrac{2t}{1+t^2}$ **2** $\tan 2A = \dfrac{1+t^2}{1-t^2}$ **3** $\cos 2A = \dfrac{1-t^2}{1+t^2}$

$$\sin 2A = 2\sin A\cos A = \frac{2\sin A\cos A}{\cos^2 A + \sin^2 A} = \frac{2\tan A}{1+\tan^2 A}$$

(Dividing top and bottom by $\cos^2 A$) so statement 1 is correct

$$\cos 2A = \cos^2 A - \sin^2 A = \frac{\cos^2 A - \sin^2 A}{\cos^2 A + \sin^2 A} = \frac{1-\tan^2 A}{1+\tan^2 A}$$

so statement 3 is correct

$$\tan 2A = \frac{\sin 2A}{\cos 2A} = \frac{2t}{1-t^2}$$

so statement 2 is wrong

ANSWER B

Example 8
$\cos 2A =$

1 $\cos^2 A + \sin^2 A$ **2** $2\cos^2 A - 1$ **3** $1 - 2\sin^2 A$

$\cos(A+B) = \cos A \cos B - \sin A \sin B$ and putting $B = A$ gives

$\cos 2A = \cos^2 A - \sin^2 A$	so statement 1 is wrong
$= 1 - \sin^2 A - \sin^2 A$	since $\cos^2 A = 1 - \sin^2 A$
$= 1 - 2\sin^2 A$	so statement 3 is correct
$= 1 - 2(1 - \cos^2 A)$	since $\sin^2 A = 1 - \cos^2 A$
$= 2\cos^2 A - 1$	so statement 2 is correct

ANSWER C

Example 9
If $180° < A < 270°$ then

1 $\sin A$ is positive **2** $\cos A$ is positive
3 $\tan A$ is positive

In the third quadrant $\tan A$ is the only ratio which is positive. Most students devise a way for remembering which ratios are positive in the four quadrants.

For $0° < A < 90°$ in the first quadrant **All** ratios are **Positive**

For $90° < A < 180°$ in the second quadrant the **Sine** is **Positive**

For $180° < A < 270°$ in the third quadrant the **Tangent** is **Positive**

For $270° < A < 360°$ in the fourth quadrant the **Cosine** is **Positive**

Example 10
A curve is given parametrically by $x = \cos t$, $y = \cos 2t$

1 The curve is symmetrical about the y-axis.
2 The curve is a parabola.
3 Its Cartesian equation is $y = 2x^2 - 1$.

Consider statement 3 first: $y = \cos 2t = 2\cos^2 t - 1 = 2x^2 - 1$ which is symmetrical about the y-axis, but the values for x and y lie between -1 and $+1$ so the curve is that part of the parabola lying within the square $-1 \leqslant x \leqslant +1$ and $-1 \leqslant y \leqslant +1$. Statement 2 is therefore only partly true. ANSWER B

Example 11
$I = \int 2 \sin x \cos x \, dx =$

1 $\sin^2 x + c$ **2** $-\cos^2 x + k$ **3** $-\frac{1}{2}\cos 2x + K$

Substituting $y = \sin x$ gives $dy = \cos x \, dx$ so

$$I = \int 2y \, dy = y^2 + c = \sin^2 x + c$$

statement 1 is correct

Substituting $y = \cos x$ gives $dy = -\sin x \, dx$ so

$$I = \int -2y \, dy = -y^2 + k = -\cos^2 x + k$$

statement 2 is correct

$$I = \int \sin 2x \, dx = -\tfrac{1}{2}\cos 2x + K$$

statement 3 is correct

All three answers are correct. ANSWER A

At first these results puzzle the student but the three answers only differ from each other by a constant amount.

$\sin^2 x = 1 - \cos^2 x$ and $-\frac{1}{2}\cos 2x = -\frac{1}{2}(2\cos^2 x - 1) = \frac{1}{2} - \cos^2 x$

When the limits of integration are inserted all three give the same result.

Example 12
In comparison with $y = \sin x$ the graph of

1 $y = 2 \sin x$ has double the frequency.
2 $y = \sin 2x$ has double the amplitude.
3 $y = 2 \sin 2x$ has double the amplitude and double the frequency.

In statement 1, each value of $\sin x$ is doubled so the graph will appear twice as high, i.e., it will take values between -2 and $+2$ while $y = \sin x$ lies between -1 and $+1$. Its amplitude is doubled. $y = \sin x$ completes a full cycle while x changes by 2π radians (360°) whereas $2x$ will complete a full cycle while x changes by π radians (180°). So $\sin 2x$ completes two cycles while $\sin x$ is completing only one and so has double the frequency. Statements 1 and 2 are both wrong, being the wrong way round, and statement 3 is correct.
ANSWER E

Example 13
The factors of $\sin 3A + \sin 5A$ are

1 $\sin A$ and $\sin 4A$ **2** $\sin A$ and $\cos 4A$ **3** $\cos A$ and $\sin 4A$

The required factor formula is
$\sin P + \sin Q = 2 \sin \frac{1}{2}(P+Q) \cos \frac{1}{2}(P-Q)$ which gives

$$\begin{aligned}\sin 3A + \sin 5A &= 2 \sin \tfrac{1}{2}(3A+5A) \cos \tfrac{1}{2}(3A-5A)\\ &= 2 \sin 4A \cos(-A)\\ &= 2 \sin 4A \cos A\end{aligned}$$

So the factors are in statement 3. ANSWER E

Example 14
The graph of $y = \cos^2 x$

1 is periodic **2** has amplitude 1 **3** is symmetrical

$\cos 2x = 2\cos^2 x - 1 \Rightarrow \cos^2 x = \frac{1}{2}(1 + \cos 2x)$ which has period π radians (180°) and varies between 0 and 1 and so has amplitude $\frac{1}{2}$. Being a transformation of $\cos x$ it has vertical lines of symmetry and rotational symmetry. So statements 1 and 3 are correct.
ANSWER B

General questions

Example 15
Expand $\sin x \cos x$ as a series in ascending powers of x as far as the term in x^3. Compare this with the series for $\sin 2x$.

$\sin x = x - \frac{1}{6}x^3 + \ldots$ and $\cos x = 1 - \frac{1}{2}x^2 + \ldots$

$\sin x \cos x = (x - \frac{1}{6}x^3)(1 - \frac{1}{2}x^2) = x - \frac{1}{2}x^3 - \frac{1}{6}x^3 + \ldots = x - \frac{2}{3}x^3$

$$\sin 2x = 2x - \frac{(2x)^3}{3!} = 2x - \frac{4x^3}{3}$$

This is double the series of $\sin x$ which is just what we expect as $\sin 2x = 2 \sin x \cos x$.

> **Example 16**
> Given that $\tan x = \frac{5}{12}$ and $0° < x < 90°$ find $\tan 2x$, $\sin 2x$ and $\cos 2x$.

$\tan x = \frac{5}{12} \Rightarrow \sin x = \frac{5}{13}$ and $\cos x = \frac{12}{13}$. These values can be found easily by drawing a right-angled triangle, putting 5 on the side opposite angle x and 12 adjacent and giving a hypotenuse of 13 (by Pythagoras). The sine and cosine ratios can then be calculated. This process is used quite often, especially with 3,4,5 and 5,12,13 triangles.

$$\tan 2x = \frac{2\tan x}{1-\tan x} = \frac{\frac{5}{6}}{1-\frac{25}{144}} = \frac{\frac{5}{6}}{\frac{119}{144}} = \frac{120}{119}$$

$$\sin 2x = 2\sin x\cos x = 2\times\frac{5}{13}\times\frac{12}{13} = \frac{120}{169}$$

$$\cos 2x = \cos x - \sin x = \frac{144}{169} - \frac{25}{169} = \frac{119}{169}$$

Incidentally we have found another Pythagorean triple 119,120,169 making a right-angled triangle with integer (whole number) sides, whose angles are very close to 45°, 45°, 90°.

> **Example 17**
> Given that $\cos 2x = \frac{119}{169}$ and $0° < x < 90°$ find the value of $\cos x$ without using tables or a calculator.

$$\cos 2x = 2\cos^2 x - 1 = \frac{119}{169} \Rightarrow 2\cos^2 x = 1 + \frac{119}{169} = \frac{288}{169}$$

$$\Rightarrow \cos^2 x = \frac{144}{169} \Rightarrow \cos x = \frac{12}{13}$$

This can be seen from Example 16 to be correct.

> **Example 18**
> Without using tables find the values of sin 15°, cos 15° and tan 15°, leaving your answers in surd form.

$$\sin 15° = \sin(45° - 30°) = \sin 45°\cos 30° - \cos 45°\sin 30°$$

$$= \frac{1}{\sqrt{2}}\times\frac{\sqrt{3}}{2} - \frac{1}{\sqrt{2}}\times\frac{1}{2} = \frac{\sqrt{3}-1}{2\sqrt{2}}$$

$$\cos 15° = \cos(45° - 30°) = \cos 45°\cos 30° + \sin 45°\sin 30°$$

$$= \frac{1}{\sqrt{2}}\times\frac{\sqrt{3}}{2} + \frac{1}{\sqrt{2}}\times\frac{1}{2} = \frac{\sqrt{3}+1}{2\sqrt{2}}$$

$$\tan 15° = \frac{\sin 15°}{\cos 15°} = \frac{\sqrt{3}-1}{\sqrt{3}+1}$$

Example 19
Find the mean value of the function $\sin x$ over the interval $0 < x < \dfrac{\pi}{2}$.

The mean value of a function is given by the integral over that interval divided by the length of the interval.

$$\text{Mean value} = \frac{\int \sin x\,dx}{\frac{1}{2}\pi} = \frac{2}{\pi}[-\cos x]_0^{\pi/2} = \frac{2}{\pi}\left(-\cos\frac{\pi}{2} + \cos 0\right) = \frac{2}{\pi}.$$

It is worth remembering that the area under the graph of $\sin x$ or $\cos x$ from 0 to $\pi/2$ is 1.

Example 20
Find the mean value of the function $y = \cos^2 x$ for $0 \leqslant x \leqslant \pi$.

$$\text{Mean value} = \frac{1}{\pi}\int \cos^2 x\,dx$$

$$= \frac{1}{\pi}\int_0^{\pi} \tfrac{1}{2}(1 + \cos 2x)\,dx$$

$$= \frac{1}{2\pi}[x + \tfrac{1}{2}\sin 2x]_0^{\pi}$$

$$= \frac{1}{2\pi} \times \pi = \tfrac{1}{2}$$

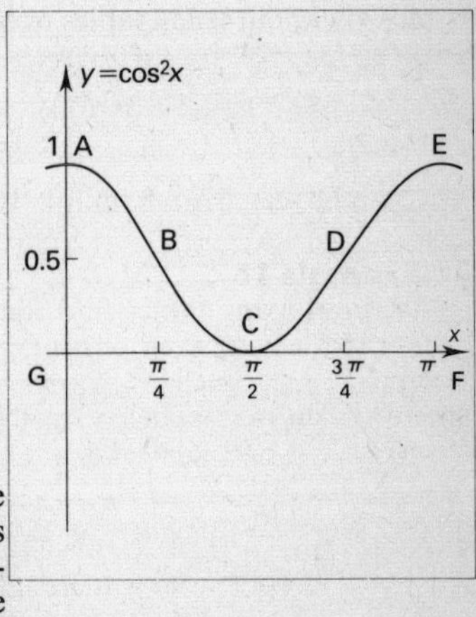

Figure 4

There is no need to calculate the integral of $\cos x$. This can be done using the symmetry of the graph in Figure 4. For every value of the function from 0° to 45° (*A* to *B*) there is a value from 45° to 90° (*B* to *C*) which complements it so that the sum of the two values is 1 and their average is $\frac{1}{2}$. This also applies to the values between *CD* and *DE*.

Example 21

A watering trough is 1 metre long and its cross-section is in the form of the graph of $y = \cos^2 x$ for $0 \leqslant x \leqslant \pi$. If the trough is 60 cm wide and 30 cm deep find its capacity.

We need to find the area of cross-section which will be similar in shape to the graph in Figure 4. The graph $ABCDE$ divides the rectangle $AEFG$ in half and the rectangle has area $AG \times GF = 1 \times \frac{\pi}{2}$.

The cross-sectional area of the trough will be $\frac{1}{2} \times 30 \times 60\,\text{cm}^2$.
Volume of trough $= \frac{1}{2} \times 30 \times 60 \times 100\,\text{cm}^3 = 90\,000$ cc $= 90$ litres.

Example 22

What are the periods of

(a) $\cos x$ (b) $\cos^2 x$ (c) $\cos^3 x$

(a) The period is the time (or length) of one oscillation, or the difference in x values between peak values (or troughs). For $\cos x$ this is 2π radians.

(b) From Figure 4 the graph of $\cos^2 x$ completes one cycle in π^c so its period is π^c.

(c) $\cos 3x = 4\cos^3 x - 3\cos x \Rightarrow \cos^3 x = \frac{3}{4}\cos x + \frac{1}{4}\cos 3x$.

Cos x has period 2π; $\cos 3x$ has period $\frac{2\pi}{3}$ and repeats its values for the third time every $2\pi^c$. Cos$^3 x$ will not repeat values until $\cos x$ does, i.e. after $2\pi^c$. Its period is $2\pi^c$.

Example 23

(a) Show that $\sin 3A = 3\sin A - 4\sin^3 A$
(b) Hence, or otherwise, solve $\sin 3A = \sin^2 A$

(a)
$$\begin{aligned}\sin 3A = \sin(2A + A) &= \sin 2A \cos A + \cos 2A \sin A \\ &= 2\sin A \cos A \cos A + (1 - 2\sin^2 A)\sin A \\ &= 2\sin A(1 - \sin^2 A) + \sin A - 2\sin^3 A \\ &= 2\sin A - 2\sin^3 A + \sin A - 2\sin^3 A \\ &= 3\sin A - 4\sin^3 A.\end{aligned}$$

(b) $\sin 3A = \sin^2 A \Rightarrow 3\sin A - 4\sin^3 A = \sin^2 A$

$$\Rightarrow 4\sin^3 A + \sin^2 A - 3\sin A = 0$$
$$\Rightarrow \sin A(4\sin^2 A + \sin A - 3) = 0$$
$$\Rightarrow \sin A(\sin A + 1)(4\sin A - 3) = 0$$

$\Rightarrow \sin A = 0$ or $\sin A = -1$ or $\sin A = \frac{3}{4}$

$\Rightarrow A = 0°, 180°, 360°$ or $A = 270°$ or $A = 48.6°, 131.4°$

> **Example 24**
> Solve the equation $3\sin A + 4\cos A = 2$ for $0° \leqslant A \leqslant 360°$

Method 1 $\frac{3}{5}\sin A + \frac{4}{5}\cos A = \frac{2}{5}$

Write $\cos B = \frac{3}{5}$ and $\sin B = \frac{4}{5} \Rightarrow B = 53.1°$

So $\cos B \sin A + \sin B \cos A = 0.4$

$\sin(A+B) = 0.4 \Rightarrow A + B = 23.6°$ or $156.4° + 360k°$

$$\begin{aligned} B = 53.1° \Rightarrow A &= 23.6° - 53.1° + 360k° \text{ or } 156.4° - 53.1° + 360k° \\ &= 360° - 29.5° + 360k° \text{ or } 103.3° + 360k° \\ &= 330.5° + 360k° \text{ or } 103.3° + 360k° \end{aligned}$$

Method 2 If $t = \tan\frac{1}{2}A$ then $\sin A = \dfrac{2t}{1+t^2}$ and $\cos A = \dfrac{1-t^2}{1+t^2}$.

Substituting these in the equation gives $6t + 4(1-t^2) = 2(1+t^2)$

$\Rightarrow 6t + 4 - 4t^2 = 2 + 2t^2$

$\Rightarrow 6t^2 - 6t - 2 = 0 \Rightarrow 3t^2 - 3t - 1 = 0$

$$\Rightarrow t = \frac{3 \pm \sqrt{9+12}}{6} = \frac{3 \pm 4.5826}{6} = 1.2638 \text{ or } -0.2638$$

$\tan\frac{1}{2}A = 1.2638 \Rightarrow \frac{1}{2}A = 51.65° \Rightarrow A = 103.3°$

$\tan\frac{1}{2}A = -0.2638 \Rightarrow \frac{1}{2}A = -14.78° \Rightarrow A = -29.5°$ or $330.5°$

General solution $\Delta = 103.3° + 360k°$ or $330.5° + 360k°$.

> **Example 25**
> Show that for all A, $\cos A + \cos(A + 120°) + \cos(A + 240°) = 0$.

Method 1 $\cos(A + 120°) = \cos A \cos 120° - \sin A \sin 120°$

$$= -\tfrac{1}{2}\cos A - \frac{\sqrt{3}}{2}\sin A$$

$$\cos(A + 240°) = \cos A \cos 240° - \sin A \sin 240° = -\tfrac{1}{2}\cos A + \frac{\sqrt{3}}{2}\sin A$$

$\cos A+\cos(A+120°)+\cos(A+240°)=0$

Method 2 Use $\cos P+\cos Q=2\cos\frac{1}{2}(P+Q)\cos\frac{1}{2}(P-Q)$ to give

$$\cos A+\cos(A+240°)=2\cos\tfrac{1}{2}(2A+240°)\cos\tfrac{1}{2}(-240°)$$
$$=-\cos(A+120°)\quad\text{since}\cos(-120°)=-\tfrac{1}{2}$$

Hence $\cos A+\cos(A+240°)+\cos(A+120°)=0$

Method 3 (See Figure 5 for vector diagram.) PQR is an equilateral triangle with PQ making an angle of A with the positive x-axis.

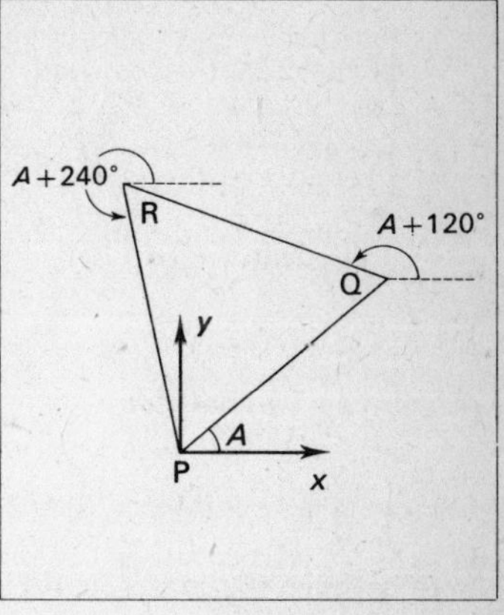

Figure 5

Vector $\mathbf{PQ}=\begin{pmatrix}\cos A\\ \sin A\end{pmatrix}$ or $\cos A\,\mathbf{i}+\sin A\,\mathbf{j}$

QR makes an angle of $(A+120°)$ with the positive x-axis, so

$$\mathbf{QR}=\begin{pmatrix}\cos(A+120°)\\ \sin(A+120°)\end{pmatrix}$$

and

$$\mathbf{RP}=\begin{pmatrix}\cos(A+240°)\\ \sin(A+240°)\end{pmatrix}$$

The vectors representing the sides of the triangle add to zero, i.e.,

$\mathbf{PQ}+\mathbf{QR}+\mathbf{RP}=\mathbf{0}\Rightarrow$

$$\begin{pmatrix}\cos A\\ \sin A\end{pmatrix}+\begin{pmatrix}\cos(A+120°)\\ \sin(A+120°)\end{pmatrix}+\begin{pmatrix}\cos(A+240°)\\ \sin(A+240°)\end{pmatrix}=\begin{pmatrix}0\\ 0\end{pmatrix}$$

This gives the required result and a similar result for sines.
In this example, the student will probably find Method 1 the easiest but Method 2 illustrates how useful the factor formulae can be and Method 3 shows how vectors can be used. Use of the methods of matrices and vectors is the best way of deriving the addition formulae used in Method 1.

Example 26
On the same diagram sketch the curves whose polar equations are

$C\quad r=1+\cos\theta,\qquad C'\quad r=2+\cos\theta$

Find the area enclosed between the two curves.

Both curves are symmetrical about the x-axis (see Figure 6(a)). The area enclosed by a curve is given by $\int \frac{1}{2}r^2\theta\, d\theta$.

For C,

$$A = 2\int_0^{\pi} \tfrac{1}{2}(1+\cos\theta)^2 d\theta$$

$$= \int_0^{\pi} (1+2\cos\theta+\cos^2\theta)d\theta$$

$$= \int_0^{\pi} (1+2\cos\theta+\tfrac{1}{2}(1+\cos 2\theta))d\theta$$

$$= [1\tfrac{1}{2}\theta+2\sin\theta+\tfrac{1}{4}\sin 2\theta]_0^{\pi} = 1\tfrac{1}{2}\pi \text{ since } \sin\pi = \sin 2\pi = \sin 0 = 0.$$

For C',

$$A = 2\int_0^{\pi} \tfrac{1}{2}(2+\cos\theta)^2 d\theta$$

$$= \int_0^{\pi} (4+4\cos\theta+\cos^2\theta)d\theta = 4\tfrac{1}{2}\pi$$

Area between the curves $= 4\frac{1}{2}\pi - 1\frac{1}{2}\pi = 3\pi$.

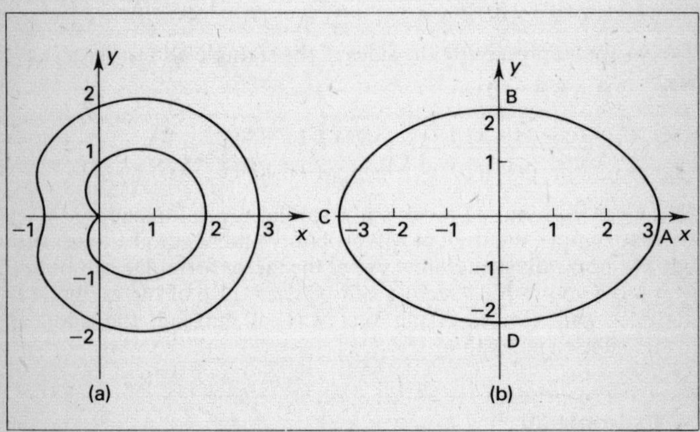

Figure 6

As a check on these two integrals C is a little larger than the circle centre (1,0) radius 1 which has area π, and C' is a little larger than the circle centre (1,0) radius 2 which has area 4π.

Chapter 3
Differentiation

Differentiation is the name given to the process by which the rate of change of one variable with respect to another can be found.

Clearly it is important that the basic concepts of differentiation should be thoroughly understood before attempting to apply the ideas to particular problems.

There are a number of fundamental methods and techniques for evaluating the differentials of more complex functions. Those required by most syllabuses include the product and quotient rules, the function of a function method, implicit and parametric differentiation and logarithmic differentiation.

These basic techniques can be used on algebraic, trigonometrical, exponential, logarithmic and inverse functions. Questions may be set which simply test these processes or their application to problems such as turning points of a curve, velocity and acceleration, etc.

Multiple choice questions (Type A) Select the correct answer

Example 1

If $y = x/\sin x$ then $\dfrac{dy}{dx} =$

A $\dfrac{\sin x - x\cos x}{x^2}$ **B** $\dfrac{\sin x + x\cos x}{\sin^2 x}$ **C** $\dfrac{\sin x - x\cos x}{\sin^2 x}$

D $\dfrac{x\cos x - \sin x}{\sin^2 x}$ **E** $\dfrac{x\cos x - \sin x}{x^2}$

Since the given function is a quotient in the form $y = \dfrac{u}{v}$ the solution can be found by using the quotient rule.

$$\frac{d}{dx}\left(\frac{u}{v}\right) = \frac{v\left(\dfrac{du}{dx}\right) - u\left(\dfrac{dv}{dx}\right)}{v^2}$$

Thus $u = x \Rightarrow \dfrac{du}{dx} = 1$ and $v = \sin x \Rightarrow \dfrac{dv}{dx} = \cos x$

If $y=\dfrac{x}{\sin x}$ then $\dfrac{dy}{dx}=\dfrac{1\times\sin x-x\times\cos x}{\sin^2 x}=\dfrac{\sin x-x\cos x}{\sin^2 x}$

ANSWER C

Example 2

If $y=\ln\left(\cos\dfrac{x}{3}\right)$ then $\dfrac{dy}{dx}=$

A $-\frac{1}{3}\tan\left(\dfrac{x}{3}\right)$ **B** $-\tan\left(\dfrac{x}{3}\right)$ **C** $-\dfrac{1}{\cos\left(\dfrac{x}{3}\right)}$

D $-\dfrac{1}{3\cos\left(\dfrac{x}{3}\right)}$ **E** $-\dfrac{3}{\cos\left(\dfrac{x}{3}\right)}$

If we let $u=\cos\left(\dfrac{x}{3}\right)$ then $y=\ln u$.

Hence y is a function of u which is itself a function of x and thus the function of a function result can be used.

If $u=\cos\left(\dfrac{x}{3}\right)$ then $\dfrac{du}{dx}=-\frac{1}{3}\sin\left(\dfrac{x}{3}\right)$

If $y=\ln u$ then $\dfrac{dy}{du}=\dfrac{1}{u}=\dfrac{1}{\cos\left(\dfrac{x}{3}\right)}$

As $\dfrac{dy}{dx}=\dfrac{dy}{du}\times\dfrac{du}{dx}$ we have $\dfrac{dy}{dx}=-\frac{1}{3}\sin\left(\dfrac{x}{3}\right)\times\dfrac{1}{\cos\left(\dfrac{x}{3}\right)}$

$$=-\tfrac{1}{3}\tan\left(\frac{x}{3}\right)$$

ANSWER A

Example 3

$\lim\limits_{x\to\infty}\left(\dfrac{3x^2+x-2}{x^2-x+1}\right)=$

A 0 **B** 1 **C** 2 **D** 3 **E** ∞

In this example it is best to rearrange the function by dividing the numerator by the denominator to give

$$\frac{3x^2+x-2}{x^2-x+1} = 3+\frac{4x-5}{x^2-x+1}$$

Thus $\lim_{x\to\infty}\left(\frac{3x^2+x-2}{x^2-x+1}\right) = \lim_{x\to\infty}\left(3+\frac{4x-5}{x^2-x+1}\right) = 3,$

since the term $\frac{4x-5}{x^2-x+1}\to 0$ as $x\to\infty$. ANSWER D

Alternatively, by dividing the numerator and denominator by x^2

$$\lim_{x\to\infty}\left(\frac{3x^2+x-2}{x^2-x+1}\right) = \lim_{x\to\infty}\left(\frac{3+\frac{1}{x}-\frac{2}{x^2}}{1-\frac{1}{x}+\frac{1}{x^2}}\right) = 3$$

Multiple choice questions (Type B) Answer according to the table

A	B	C	D	E
1, 2, 3 correct	1, 3 only	2, 3 only	2 only	3 only

Example 4
If $x = 3\cos\theta$ and $y = 3\sin\theta$

1 The point $P(x,y)$ lies on a circle of radius 3.
2 The gradient of the curve at P is $-\cot\theta$.
3 The normal at P has the equation $y = x\tan\theta$.

Since $x = 3\cos\theta$ and $y = 3\sin\theta$ (1)

$\cos\theta = \frac{x}{3}$ and $\sin\theta = \frac{y}{3}$

As $\cos^2\theta+\sin^2\theta = 1$ we have $\left(\frac{x}{3}\right)^2+\left(\frac{y}{3}\right)^2 = 1.$

Thus, eliminating the parameter θ has produced the Cartesian equation of the curve, i.e.

$$x^2+y^2 = 9$$

which represents the equation of a circle of radius 3.

From (1) it follows that $\dfrac{dx}{d\theta} = -3\sin\theta$ and $\dfrac{dy}{d\theta} = 3\cos\theta$.

Using the function of a function result $\dfrac{dy}{dx} = \dfrac{dy}{d\theta} \times \dfrac{d\theta}{dx}$

$$\frac{dy}{dx} = -\frac{3\cos\theta}{3\sin\theta} = -\cot\theta \qquad (2)$$

Thus the gradient of the tangent at P is $-\cot\theta$.

From (2) it follows that the gradient of the normal to the curve at P is $\tan\theta$ (the product of the gradients of perpendicular lines is -1).

The equation of the normal at P is

$$y - 3\sin\theta = \tan\theta\,(x - 3\cos\theta)$$

$$y - 3\sin\theta = x\tan\theta - 3\sin\theta$$

$$y = x\tan\theta \qquad (3)$$

Statements 1, 2 and 3 are all true. ANSWER A

General questions

In questions that require written solutions the answer has to be evaluated and attention may be confined to one particular topic on the syllabus or involve several different aspects. The questions will test a basic method or an application of differentiation.

Example 5

Differentiate the function $x^2 - 2x - 3$ with respect to x from first principles.

Consider the curve $y = x^2 - 2x - 3$ as shown in Figure 7. Let $P(x,y)$ be a general point of the curve and $Q(x+\delta x, y+\delta y)$ a neighbouring point.

Since Q lies on the curve, the coordinates $(x+\delta x, y+\delta y)$ satisfy the equation of the curve.

$$y + \delta y = (x+\delta x)^2 - 2(x+\delta x) - 3 \qquad (1)$$

Also, since $P(x,y)$ lies on the curve

$$y = x^2 - 2x - 3 \qquad (2)$$

Subtracting (2) from (1)

$$\begin{aligned}\delta y &= (x+\delta x)^2 - 2(x+\delta x) - 3 - (x^2 - 2x - 3)\\ &= x^2 + 2x(\delta x) + (\delta x)^2 - 2x - 2(\delta x) - 3 - x^2 + 2x + 3\\ &= 2x(\delta x) - 2(\delta x) + (\delta x)^2\end{aligned}$$

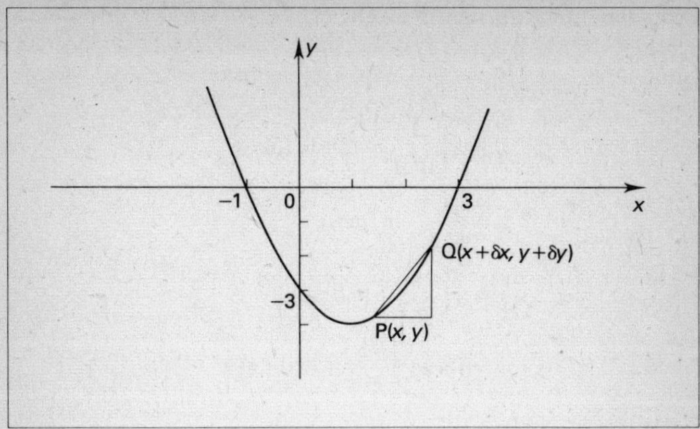

Figure 7

$$\text{The gradient of the chord } PQ = \frac{\delta y}{\delta x} = \frac{2x(\delta x) - 2(\delta x) + (\delta x)^2}{\delta x}$$

$$= 2x - 2 + (\delta x)$$

Now the gradient of the curve at a point is defined as the gradient of the tangent at that point. This is the limit of the gradient of the chord PQ as $\delta x \to 0$.

$$\text{Hence } \frac{dy}{dx} = \lim_{\delta x \to 0} \left(\frac{\delta y}{\delta x} \right) = \lim_{\delta x \to 0} [2x - 2 + (\delta x)]$$

$$\frac{dy}{dx} = 2x - 2$$

Example 6
Differentiate

(a) $(\ln x)^x$ (b) $\dfrac{(1+x^2)(2-x)^3}{(1+x^3)^2}$ with respect to x.

(a) Let $y = (\ln x)^x$
Taking logarithms of both sides we have

$$\ln y = \ln (\ln x)^x = x \ln (\ln x)$$

Differentiating with respect to x

$$\frac{1}{y}\frac{dy}{dx} = x\frac{d}{dx}[\ln(\ln x)] + \ln(\ln x)$$

$$= x\left[\frac{1}{\ln x}\times\frac{1}{x}\right] + \ln(\ln x)$$

$$= \frac{1}{\ln x} + \ln(\ln x)$$

$$\frac{dy}{dx} = y\left[\frac{1}{\ln x} + \ln(\ln x)\right]$$

$$= (\ln x)^x\left[\frac{1}{\ln x} + \ln(\ln x)\right]$$

$$\frac{dy}{dx} = (\ln x)^{x-1} + (\ln x)^x \ln(\ln x)$$

(b) The idea of taking logarithms can also be used for a function of this form instead of the product and quotient rules.

Let $y = \dfrac{(1+x^2)(2-x)^3}{(1+x^3)^2}$

Taking logarithms of both sides we have

$$\ln y = \ln(1+x^2) + \ln(2-x)^3 - \ln(1+x^3)^2$$

$$\Rightarrow \ln y = \ln(1+x^2) + 3\ln(2-x) - 2\ln(1+x^3)$$

Differentiating with respect to x

$$\frac{1}{y}\frac{dy}{dx} = \frac{2x}{(1+x^2)} - \frac{3}{(2-x)} - \frac{6x^2}{(1+x^3)}$$

$$\frac{1}{y}\frac{dy}{dx} = \frac{2x(2-x)(1+x^3) - 3(1+x^2)(1+x^3) - 6x^2(1+x^2)(2-x)}{(1+x^2)(2-x)(1+x^3)}$$

$$\frac{1}{y}\frac{dy}{dx} = \frac{x^5 - 8x^4 + 3x^3 - 17x^2 + x^2 + 4x - 3}{(1+x^2)(2-x)(1+x^3)}$$

$$\therefore \frac{dy}{dx} = \frac{(1+x^2)(2-x)^3}{(1+x^3)^2}\left[\frac{x^5 - 8x^4 + 3x^3 - 17x^2 + 4x - 3}{(1+x^2)(2-x)(1+x^3)}\right]$$

$$\therefore \frac{dy}{dx} = \frac{(2-x)^2}{(1+x^3)^3}(x^5 - 8x^4 + 3x^3 - 17x^2 + 4x - 3)$$

Note Remember that logarithms are taken to base e (written $\ln x$)

in order that $\frac{d}{dx}(\ln x) = \frac{1}{x}$. These are called natural logarithms.

> **Example 7**
> A particle moves in a straight line in such a way that its displacement from a fixed point O, t seconds after passing that point, is given by $s = 6\sin t + 3\cos 2t$ metres. Find
>
> (a) the time when the particle is first at rest,
> (b) its displacement from O at that instant,
> (c) its acceleration at that instant.

(a) Since velocity = rate of change of displacement

$$\text{velocity} = \frac{ds}{dt} = 6\cos t - 6\sin 2t$$

Thus, when the velocity is zero, $\frac{ds}{dt} = 0$.

$$6\cos t - 6\sin 2t = 0$$
$$\cos t - 2\sin t\cos t = 0$$
$$\cos t\,(1 - 2\sin t) = 0$$

This gives $\cos t = 0$ or $\sin t = \frac{1}{2}$.

If $\cos t = 0$, $t = \frac{\pi}{2}, \frac{3\pi}{2}, \ldots$

If $\sin t = 0$, $t = \frac{\pi}{6}, \frac{5\pi}{6}, \ldots$

Thus the particle is first at rest when $t = \frac{\pi}{6}$ seconds.

(b) The displacement from O is given by $s = 6\sin t + 3\cos 2t$.

When $t = \frac{\pi}{6}$ $\quad s = 6\sin\frac{\pi}{6} + 3\cos\frac{\pi}{3}$

$$= 3 + 1\tfrac{1}{2} = 4\tfrac{1}{2}\text{ m}$$

(c) Since acceleration = rate of change of velocity

$$\text{Acceleration} = \frac{dv}{dt} = \frac{d}{dt}(6\cos t - 6\sin 2t)$$

Thus $\quad \text{acceleration} = -6\sin t - 12\cos 2t$

When $t = \frac{\pi}{6}$ acceleration $= -6\sin\frac{\pi}{6} - 12\cos\frac{\pi}{3}$

$$= -3 - 6 = -9\,\text{m}\,\text{s}^{-2}$$

Note

1. There are many variations to this type of question and integration is often also included. The solutions require knowledge of the following results

$$v = \frac{ds}{dt} \quad \text{or} \quad s = \int v\,dt$$

$$a = \frac{dv}{dt} \quad \text{or} \quad v = \int a\,dt$$

where s is the displacement, v is the velocity and a is the acceleration of the particle.

2. Remember that the solutions of any equation in t must be in radians and not in degrees.

Example 8

If $y = \dfrac{2\cos nx}{x}$ where n is a constant prove that

$$x\frac{d^2y}{dx^2} + 2\frac{dy}{dx} + n^2xy = 0$$

Since $y = \dfrac{2\cos nx}{x}$ it follows that $xy = 2\cos nx$.

Differentiating implicitly with respect to x

$$x\frac{dy}{dx} + y = -2n\sin nx$$

Differentiating implicitly a second time with respect to x

$$x\frac{d^2y}{dx^2} + \frac{dy}{dx} + \frac{dy}{dx} = -2n^2\cos nx$$

$$x\frac{d^2y}{dx^2} + 2\frac{dy}{dx} = -n^2xy$$

or

$$x\frac{d^2y}{dx^2} + 2\frac{dy}{dx} + n^2xy = 0$$

Note This solution illustrates the power of implicit differentiation since the alternative method of forming the first and second differentials separately is more lengthy and cumbersome.

Example 9
Find the coordinates of any turning points or points of inflexion on the curve $y = x^4 - 4x^3$. Sketch the curve.

The gradient of the curve is given by $\dfrac{dy}{dx} = 4x^3 - 12x^2$.

Turning points occur when $\dfrac{dy}{dx} = 0$, i.e.

when
$$4x^3 - 12x^2 = 0$$
$$4x^2(x-3) = 0$$
$$x = 0 \text{ or } x = 3$$

If $x = 0$, $y = 0$.

If $x = 3$, $y = (3)^4 - 4(3)^3 = -27$.

Thus the gradient of the curve is zero at the points (0,0) and (3, −27). The nature of the turning points can be determined by considering the second differential.

$$\frac{d^2y}{dx^2} = 12x^2 - 24x = 12x(x-2)$$

When $x = 0$, $\dfrac{d^2y}{dx^2} = 0$ but $\dfrac{d^3y}{dx^3} = 24x - 24 \neq 0$.

Thus the point (0,0) is a point of inflexion.

When $x = 3$, $\dfrac{d^2y}{dx^2} > 0$.

Thus the point (3, −27) is a minimum point.
The curve has a point of inflexion at (0,0) and a minimum at (3, −27).

Also, since when $y = 0$, $x = 0$ or $x = 4$, the curve cuts the x-axis at the points (0,0) and (4,0).

As $x \to \pm\infty$, $y \to \infty$.

The curve is sketched in Figure 8.

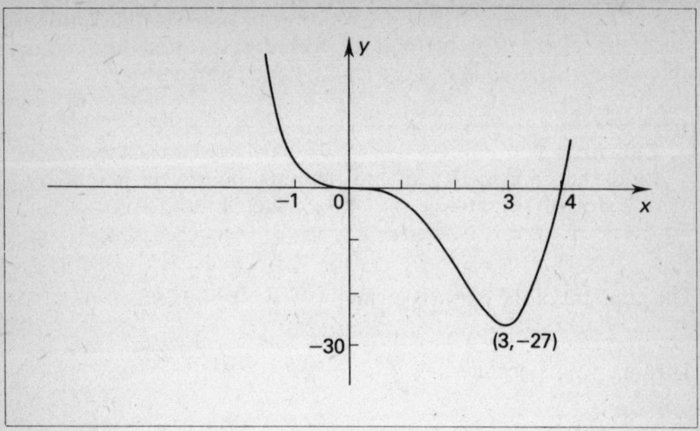

Figure 8

Note

1. Remember that a point of inflexion does not necessarily require $\frac{dy}{dx} = 0$, although it is in this case.

2. The nature of the turning points can also be determined by considering the gradient of the curve on either side of the point.

x	$x < 0$	0	$x > 0$
$\frac{dy}{dx}$	$-$ve	0	$-$ve

x	$x < 3$	3	$x > 3$
$\frac{dy}{dx}$	$-$ve	0	$+$ve

Clearly this indicates a point of inflexion at $(3, -27)$ and a minimum at $(0,0)$.

Example 10

A right circular cone is inscribed in a solid sphere of radius R. If x is the distance of the base of the cone from the centre of the sphere, prove that the volume of the cone is given by $V = \frac{1}{3}\pi(R^3 + R^2x - Rx^2 - x^3)$.

Hence find the height of the cone when the volume of the cone is a maximum. Find also, the maximum volume of the cone.

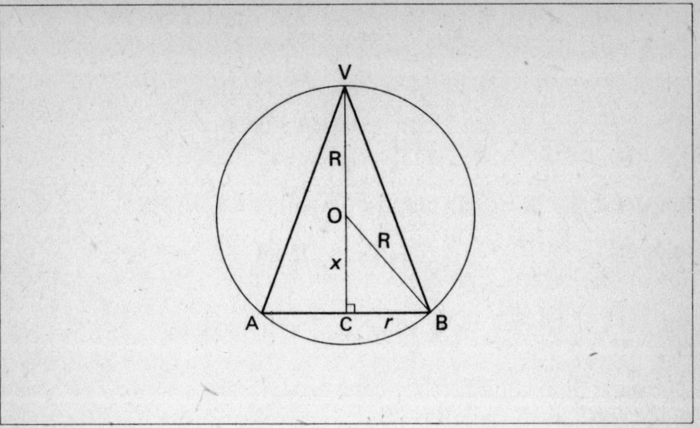

Figure 9

Figure 9 shows a cross-section of the cone VAB inscribed in a sphere, centre O and radius R.

Since $OV = OB = R$ (radius of the sphere) and $OC = x$, it follows from Pythagoras' theorem that

$$\begin{aligned} CB^2 &= OB^2 - OC^2 \\ &= R^2 - x^2 \end{aligned}$$

$\therefore$ the base radius of the cone, r, is given by

$$r = \sqrt{R^2 - x^2}$$

The volume of a cone $= \frac{1}{3}\pi r^2 h$ where r is the base radius and h is the perpendicular height.

Thus the volume of the cone, V, is given by

$$\begin{aligned} V &= \tfrac{1}{3}\pi(R^2 - x^2)(R + x) \\ &= \tfrac{1}{3}\pi(R^3 + R^2x - Rx^2 - x^3) \end{aligned}$$

For a maximum volume

$$\frac{dV}{dx} = 0 \text{ and } \frac{d^2V}{dx^2} < 0$$

Since $\qquad V = \frac{1}{3}\pi(R^3 + R^2x - Rx^2 - x^3)$

$$\frac{dV}{dx} = \tfrac{1}{3}\pi(R^2 - 2Rx - 3x^2) \qquad (R \text{ is constant})$$

When $\frac{dV}{dx} = 0$ $\qquad \frac{1}{3}\pi(R^2 - 2Rx - 3x^2) = 0$

$$R^2 - 2Rx - 3x^2 = 0$$

$$(R - 3x)(R + x) = 0$$

$$\therefore \qquad x = \tfrac{1}{3}R \text{ or } x = -R$$

Since $0 < x < R$ we can discard the solution $x = -R$.

Now $$\frac{d^2V}{dx^2} = \tfrac{1}{3}\pi(-2R - 6x)$$

Thus if $x = \frac{1}{3}R$, $\frac{d^2V}{dx^2} < 0$ which defines a maximum.

The maximum volume of the cone occurs when $x = \frac{1}{3}R$. The height of the cone when $x = \frac{1}{3}R$ is $\frac{4}{3}R$.

$\therefore$ the maximum volume of the cone is,

$$V = \tfrac{1}{3}\pi\left(R^2 - \frac{R^2}{9}\right)\left(R + \frac{R}{3}\right)$$

$$= \tfrac{1}{3}\pi\left(\frac{8R^2}{9}\right)\left(\frac{4R}{3}\right) = \frac{32\pi R^3}{81}$$

Note

1. The solution $x = -R$ does, in fact, give the minimum value that V would take if it was considered simply as a function of x without the physical limitations of the problem.

Note that when $x = -R$, $\frac{d^2V}{dx^2} > 0$.

2. Remember that the expression for the volume, V, must only contain one variable and if this is not immediately the case then steps must be taken to eliminate the extra variables before differentiating.

Chapter 4
Integration

Integration can be considered as the reverse of differentiation or as a limit of a summation of small parts. As the former of these two ideas gives an indefinite answer it is called indefinite integration, e.g.

If $y = f(x)$ and $f'(x) = x^2$ then $y = \frac{1}{3}x^3 + c$

where c is an arbitrary constant which can only be determined if further boundary conditions are given.

However, the process of adding a large number of small parts to form a sum, as in calculating an area, gives rise to a definite numerical result and is called definite integration.

There are a great number of special methods for integrating a variety of functions but most syllabuses will expect a knowledge of inspection methods, substitution and integration by parts. These will be applied to algebraic, trigonometrical, exponential and inverse functions.

Applications are varied but generally include finding areas, volumes and centroids of plane figures or solids, mean values, areas of sectors and problems involving velocity and acceleration.

Multiple choice questions (Type A) Select the correct answer

Example 1
$\int \sin^2 x \, dx =$

A $\frac{1}{3}\sin^3 x + c$ **B** $2 \sin x \cos x + c$ **C** $\frac{1}{2}x + \frac{1}{4}\cos 2x + c$
D $\frac{1}{2}x - \frac{1}{4}\sin 2x + c$ **E** $\frac{1}{2}x + \frac{1}{2}\sin 2x + c$

Since the powers of trigonometrical functions cannot be integrated immediately it is necessary to transform the integrand into a trigonometrical function of multiple angles.

Since $\sin^2 x = \frac{1}{2}(1 - \cos 2x)$ (Double angle formula)

$$\int \sin^2 x \, dx = \int (\tfrac{1}{2} - \tfrac{1}{2}\cos 2x) \, dx$$
$$= \tfrac{1}{2}x - \tfrac{1}{4}\sin 2x + c$$

ANSWER D

Example 2

Figure 10 shows the curve $y = \ln x$. The volume of the solid of revolution formed when the shaded area is rotated about the line $x = 1$ is given by

A $\pi \int_1^2 (1 - \ln x)\,dx$ **B** $\pi \int_1^2 (1 - \ln x)^2\,dx$

C $\pi \int_0^{\ln 2} (e^y - 1)^2\,dy$ **D** $\pi \int_0^{\ln 2} (\ln x)^2\,dy$ **E** $\pi \int_0^{\ln 2} e^{2y}\,dy$

If the shaded area is to be rotated about the line $x = 1$ then an element of volume can be formed by rotating an element of area about the line $x = 1$ and summing.

The volume of the approximate cylinder parallel to the x-axis is $\pi(x-1)^2\delta y$.

These elements of volume can be summed between $y = 0$ and $y = \ln 2$ to give, as $\delta y \to 0$,

$$\text{total volume} = \lim_{y \to 0} \sum_{y=0}^{\ln 2} \pi(x-1)^2\delta y$$

$$= \pi \int_0^{\ln 2} (x-1)^2\,dy$$

$$= \pi \int_0^{\ln 2} (e^y - 1)^2\,dy$$

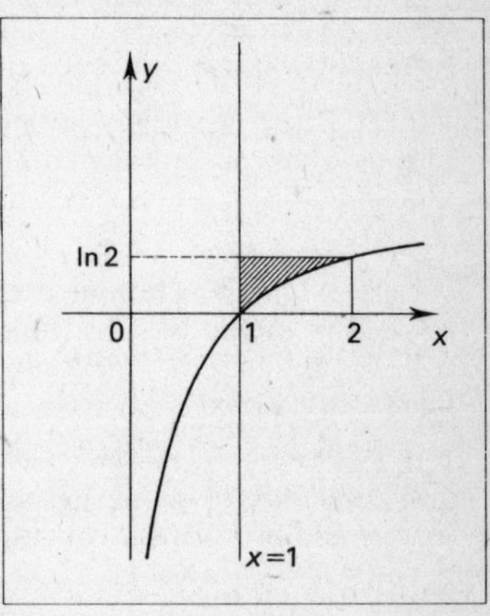

Figure 10

ANSWER C

Example 3

$\int e^{2x} \cos x\,dx =$

A $\frac{1}{5}e^{2x}(2\cos x + \sin x) + c$ **B** $\frac{1}{2}e^{2x}(\cos x + \sin x) + c$

C $\frac{1}{2}e^{2x}(\cos x - \sin x) + c$ **D** $\frac{1}{3}e^{2x}(2\cos x - \sin x) + c$

E $\frac{2}{5}e^{2x}(\cos x - 2\sin x) + c$

Since the integrand is a product we can use integration by parts.

Let $I = \int e^{2x} \cos x\,dx$

Let $u = \cos x \Rightarrow \dfrac{du}{dx} = -\sin x$ and $\dfrac{dv}{dx} = \mathrm{e}^{2x} \Rightarrow v = \frac{1}{2}\mathrm{e}^{2x}$

Since $\displaystyle\int u\frac{dv}{dx}\,dx = uv - \int v\frac{du}{dx}\,dx$ we have,

$$I = \int \mathrm{e}^{2x}\cos x\,dx = \tfrac{1}{2}\mathrm{e}^{2x}\cos x - \int(-\sin x)\tfrac{1}{2}\mathrm{e}^{2x}\,dx$$
$$= \tfrac{1}{2}\mathrm{e}^{2x}\cos x + \tfrac{1}{2}\int \mathrm{e}^{2x}\sin x\,dx$$

Repeating this process on the integral produced we have:

Let $u = \sin x \Rightarrow \dfrac{du}{dx} = \cos x$ and $\dfrac{dv}{dx} = \mathrm{e}^{2x} \Rightarrow v = \frac{1}{2}\mathrm{e}^{2x}$

$\therefore\ I = \frac{1}{2}\mathrm{e}^{2x}\cos x + \frac{1}{2}[\frac{1}{2}\mathrm{e}^{2x}\sin x - \int \cos x\,(\frac{1}{2}\mathrm{e}^{2x})\,dx]$

$\therefore\ I = \frac{1}{2}\mathrm{e}^{2x}\cos x + \frac{1}{4}\mathrm{e}^{2x}\sin x - \frac{1}{4}I$

$\therefore\ \frac{5}{4}I = \frac{1}{4}\mathrm{e}^{2x}(2\cos x + \sin x)$

or $\ I = \int \mathrm{e}^{2x}\cos x\,dx = \frac{1}{5}\mathrm{e}^{2x}(2\cos x + \sin x) + c$ ANSWER A

Note that the constant of integration is chosen to be zero when evaluating v. The incorrect answers often given in this problem stem from errors in differentiating $\cos x$ and $\sin x$ and in integrating e^{2x}. Care must be taken with the signs.

Example 4

$\displaystyle\int_0^1 3^x\,dx =$

A $\dfrac{2}{\ln 3}$ **B** 2 **C** $2\ln 3$ **D** $-\dfrac{2}{\ln 3}$ **E** no finite value

Let $3^x = u$
Taking logarithms, to base e, of both sides

$$\ln(3^x) = \ln u \Rightarrow x\ln 3 = \ln u$$

Differentiating with respect to x

$$\ln 3 = \left(\frac{1}{u}\right)\frac{du}{dx} \Rightarrow \frac{dx}{du} = \frac{1}{u\ln 3}$$

Using the method of substitution,

$$\int_0^1 3^x\,dx = \int_1^3 3^x\frac{dx}{du}\,du = \int_1^3 u\left(\frac{1}{u}\right)\left(\frac{1}{\ln 3}\right)du$$

$$\therefore \quad \int_0^1 3^x\,dx = \int_1^3 \frac{1}{\ln 3}\,du = \left[\frac{u}{\ln 3}\right]_1^3 = \frac{1}{\ln 3}(3^1 - 3^0)$$

$$\therefore \quad \int_0^1 3^x\,dx = \frac{2}{\ln 3}$$

ANSWER A

Multiple choice questions (Type B) Answer according to the table

A	**B**	**C**	**D**	**E**
1, 2, 3 correct	1, 3 only	2, 3 only	2 only	3 only

Example 5

If $f(x) = \sin x$ and $g(x) = \cos x$

1 $\displaystyle\int \frac{f(x)}{g(x)}\,dx = \ln(k\cos x)$ where k is a constant.

2 $\displaystyle\int_0^{\pi/2} f(x)\,dx = \int_0^{\pi/2} g(x)\,dx$

3 $\displaystyle\int_0^{\pi/2} \ln f(x)\,dx = \int_0^{\pi/2} \ln g(x)\,dx$

Since $\dfrac{f(x)}{g(x)} = \dfrac{\sin x}{\cos x} = \tan x$ we have

$$\int \tan x\,dx = \int \frac{\sin x}{\cos x}\,dx = -\int \frac{-\sin x}{\cos x}\,dx = -\ln(k\cos x)\cdot$$

Statement 1 is incorrect.

Now $\displaystyle\int_0^{\pi/2} f(x)\,dx = \int_0^{\pi/2} \sin x\,dx = [-\cos x]_0^{\pi/2} = 0-(-1) = 1$

Also $\displaystyle\int_0^{\pi/2} g(x)\,dx = \int_0^{\pi/2} \cos x\,dx = [\sin x]_0^{\pi/2} = 1-0 = 1$

Statement 2 is correct.

Let $x = \dfrac{\pi}{2} - u \Rightarrow \dfrac{dx}{du} = -1$

Under this transformation $x = 0 \Rightarrow u = \frac{\pi}{2}$ and $x = \frac{\pi}{2} \Rightarrow u = 0$.

$$\text{Hence} \quad \int_0^{\pi/2} \ln f(x)\,dx = \int_{x=0}^{x=\pi/2} \ln(\sin x)\frac{dx}{du}\,du$$

$$= \int_{\pi/2}^{0} \ln[\sin(\tfrac{1}{2}\pi - u)](-1)\,du$$

$$= -\int_{\pi/2}^{0} \ln(\cos u)\,du = \int_0^{\pi/2} \ln(\cos u)\,du$$

Replacing u by x we obtain

$$\int_0^{\pi/2} \ln(\sin x)\,dx = \int_0^{\pi/2} \ln(\cos x)\,dx$$

Statement 3 is correct.

Statements 2 and 3 only are true. ANSWER C

General questions

Shorter questions test a small part of the syllabus and require a basic integration method or an application of integration. Longer questions will use several aspects of the topic.

Example 6

Find the smallest positive value of n for which

$$\int_0^n 4\cos 3x \cos x\,dx = 0.$$

Using the factor formula

$$\cos P + \cos Q = 2\cos\tfrac{1}{2}(P+Q)\cos\tfrac{1}{2}(P-Q)$$

we have $\quad \tfrac{1}{2}(P+Q) = 3x \Rightarrow P+Q = 6x \qquad (1)$

and $\quad \tfrac{1}{2}(P-Q) = x \Rightarrow P-Q = 2x \qquad (2)$

Adding and subtracting equations (1) and (2) we have

$$2P = 8x \text{ and } 2Q = 4x \Rightarrow P = 4x \text{ and } Q = 2x$$

$$\therefore \quad \int_0^n 4\cos 3x \cos x\,dx = \int_0^n 2\cos 4x + 2\cos 2x\,dx$$

$$= [\tfrac{1}{2}\sin 4x + \sin 2x]_0^n = \tfrac{1}{2}\sin 4n + \sin 2n = 0$$

$$\therefore \quad \sin 2n\cos 2n + \sin 2n = \sin 2n(\cos 2n + 1) = 0$$

$$\therefore \quad \sin 2n = 0 \text{ or } \cos 2n = -1$$

If $\sin 2n = 0 \qquad 2n = 0, \pi, 2\pi, \ldots \qquad n = 0, \tfrac{1}{2}\pi, \pi, \ldots$

If $\cos 2n = -1 \qquad 2n = \pi, 3\pi, \ldots \qquad n = \tfrac{1}{2}\pi, \tfrac{3}{2}\pi, \ldots$

Thus the smallest value of $n > 0$ is $\tfrac{1}{2}\pi$.

Example 7

Evaluate $\displaystyle\int_3^4 \frac{x+4}{x^2-x-2}\,dx$

Since the denominator will factorize we use the method of partial fractions to evaluate this integral.

Let $$\frac{x+4}{x^2-x-2} \equiv \frac{x+4}{(x-2)(x+1)} \equiv \frac{A}{(x-2)} + \frac{B}{(x+1)}$$
$$\equiv \frac{A(x+1)+B(x-2)}{(x-2)(x+1)}$$

Equating the numerators
$$x+4 \equiv A(x+1)+B(x-2)$$

Let $x = 2 \Rightarrow 6 = 3A \Rightarrow A = 2$

Let $x = -1 \Rightarrow 3 = -3B \Rightarrow B = -1$

$$\therefore \quad \int_3^4 \frac{x+4}{x^2-x-2}\,dx = \int_3^4 \frac{2}{(x-2)} - \frac{1}{(x+1)}\,dx$$
$$= [2\ln(x-2) - \ln(x+1)]_3^4$$
$$= 2\ln 2 - \ln 5 + \ln 4 = \ln \tfrac{16}{5}$$

Example 8

Evaluate $\displaystyle\int \frac{dx}{x^2+6x+10}$

In this example the denominator does not factorize so we complete the square

$$\therefore \quad \int \frac{dx}{x^2+6x+10} = \int \frac{dx}{(x+3)^2+1} = \tan^{-1}(x+3)+c$$

Example 9

Evaluate $\displaystyle\int_0^{\pi/2} \frac{1}{1+\sin x}\,dx$

Use the substitution $t = \tan\frac{1}{2}x$. From the double angle formula for $\tan 2A$ we have

$$\tan x = \frac{2t}{1-t^2} \Rightarrow \cos x = \frac{1-t^2}{1+t^2} \text{ and } \sin x = \frac{2t}{1+t^2}$$

Also differentiating with respect to x

$$\frac{dt}{dx} = \tfrac{1}{2}\sec^2\frac{x}{2} = \frac{1}{2}\left(1+\tan^2\frac{x}{2}\right) = \tfrac{1}{2}(1+t^2) \Rightarrow \frac{dx}{dt} = \frac{2}{1+t^2}$$

Hence $\displaystyle\int_0^{\pi/2} \frac{1}{1+\sin x}\,dx = \int_0^1 \frac{1}{\left(1+\dfrac{2t}{1+t^2}\right)}\left(\frac{2}{(1+t^2)}\right)dt$

$$= \int_0^1 \frac{2}{1+2t+t^2}\,dt = \int_0^1 \frac{2}{(1+t)^2}\,dt = \left[\frac{-2}{1+t}\right]_0^1 = 1$$

Example 10

Evaluate $\int x\tan^{-1}x\,dx$

Integration by parts is a powerful method for integrating a product.

Let $u = \tan^{-1}x \Rightarrow \dfrac{du}{dx} = \dfrac{1}{1+x^2}$ and $\dfrac{dv}{dx} = x \Rightarrow v = \tfrac{1}{2}x^2$

Using the result $\displaystyle\int u\frac{dv}{dx}\,dx = uv - \int v\frac{du}{dx}\,dx$ we obtain

$$\begin{aligned}\int x\tan^{-1}x\,dx &= \tfrac{1}{2}x^2\tan^{-1}x - \int \tfrac{1}{2}x^2\left[\frac{1}{(1+x^2)}\right]dx \\ &= \tfrac{1}{2}x^2\tan^{-1}x - \frac{1}{2}\int \frac{x^2}{(1+x^2)}\,dx \\ &= \tfrac{1}{2}x^2\tan^{-1}x - \frac{1}{2}\int 1 - \frac{1}{(1+x^2)}\,dx \\ &= \tfrac{1}{2}x^2\tan^{-1}x - \tfrac{1}{2}x + \tfrac{1}{2}\tan^{-1}x + c\end{aligned}$$

Example 11
Find the volume of the solid formed when the region enclosed by the curve $y = \cos x$, the axes and the line $x = \pi/6$ is rotated completely about the x-axis.
Show also that Simpson's rule using 5 ordinates can be used to give a good estimate of the area under the curve.

Consider an element of volume formed by rotating an element of area about the x-axis (see Figure 11).

If $P(x, y)$ and $Q(x+\delta x, y+\delta y)$ are neighbouring points on the curve then the volume formed is $\simeq \pi y^2 \delta x$.

Summing all such elements between $x = 0$ and $x = \pi/6$.

$$\text{Total volume} \simeq \sum_{x=0}^{x=\pi/6} \pi y^2 \delta x.$$

In the limit as $\delta x \to 0$

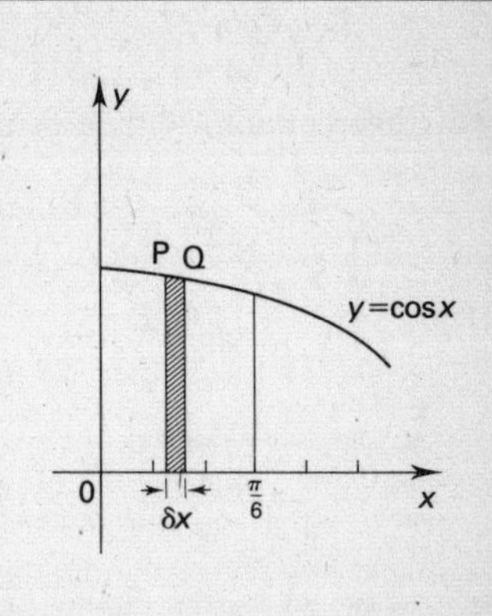

Figure 11

$$\text{volume} = \lim_{\delta x \to 0} \sum_{x=0}^{x=\pi/6} \pi y^2 \delta x$$

$$= \pi \int_0^{\pi/6} y^2\, dx = \pi \int_0^{\pi/6} \cos^2 x\, dx = \tfrac{1}{2}\pi \int_0^{\pi/6} (1+\cos 2x)\, dx$$

$$= \tfrac{1}{2}\pi[x+\tfrac{1}{2}\sin 2x]_0^{\pi/6} = \tfrac{1}{2}\pi(\pi/6+\tfrac{1}{2}\sin \pi/3)$$

$$= \pi^2/12+\sqrt{3}\pi/8$$

The volume of revolution is 1.503 to three decimal places.

Simpson's rule for 5 ordinates is given by

$$\text{Area} = \tfrac{1}{3}h(y_1+4y_2+2y_3+4y_4+y_5)$$

where $y_1, y_2, \ldots, y_5$ are the 5 ordinates.

Since the interval is 0 to $\pi/6$, $h = \pi/24$.

As $y = \cos x$

x	0	$\pi/24$	$\pi/12$	$\pi/8$	$\pi/6$
y	1	0.9914	0.9659	0.9239	0.8660

$$
\begin{array}{rrr}
y_1 = 1.0000 & y_3 = 0.9659 & y_2 = 0.9914 \\
y_5 = 0.8660 & \times 2 & y_4 = 0.9239 \\
\hline
1.8660 & 1.9318 & 1.9153 \\
1.9318 & & \times 4 \\
7.6612 & & 7.6612 \\
\hline
11.4590 & &
\end{array}
$$

$\therefore$ area under the curve $= \pi/72 \times 11.4590 = 0.500$ to three decimal places.

Note that the area is $\int_0^{\pi/6} \cos x\, dx = [\sin x]_0^{\pi/6} = 0.5$.

Example 12

Sketch the curve $y = \dfrac{6}{(2-x)(1+x)}$. Find the area bounded by the curve and (a) the line $y = 3$, (b) the x-axis between the values $x = 4$ and $x = 5$.

The following information can be deduced from the equation.

(i) $y = 3$ when $x = 0$.

(ii) As $x \to \pm\infty, y = 0$ from below.

(iii) $\dfrac{dy}{dx} = -\dfrac{6[(2-x)-(1+x)]}{(2-x)^2(1+x)^2} = \dfrac{-(6-12x)}{(2-x)^2(1+x)^2}$ (1)

Thus the gradient is zero when $x = \frac{1}{2}$. If $x < \frac{1}{2} \Rightarrow \dfrac{dy}{dx} < 0$ and if $x > \frac{1}{2} \Rightarrow \dfrac{dy}{dx} > 0$ from (1). Hence $(\frac{1}{2}, 2\frac{2}{3})$ is a minimum point of the curve.

(iv) If $x = -1$ or 2, $y \to \infty$ and these lines are asymptotes to the curve. The sketch is shown in Figure 12.

(a) The given curve meets the line $y = 3$ when

$$3(2-x)(1+x) = 6 \Rightarrow (2-x)(1+x) = 2$$
$$\Rightarrow x^2 - x = 0 \Rightarrow x(x-1) = 0 \Rightarrow x = 0 \text{ or } x = 1$$

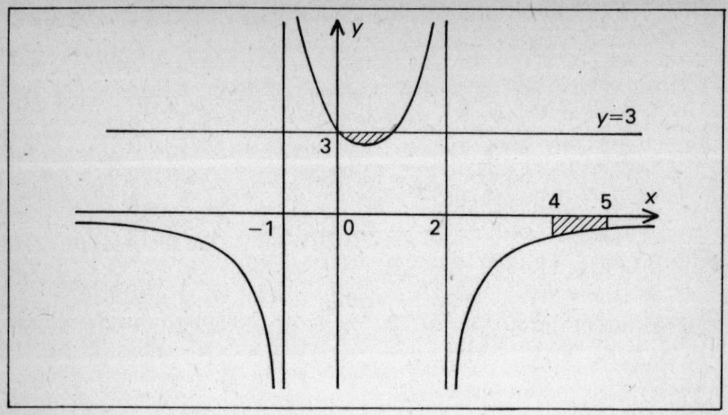

Figure 12

$$\text{Hence the required area} = \int_0^1 (3-y)\,dx$$

$$= \int_0^1 3\,dx - \int_0^1 \frac{6}{(2-x)(1+x)}\,dx$$

The integrand of the second integral can be rearranged by partial fractions.

$$\frac{6}{(2-x)(1+x)} \equiv \frac{A}{(2-x)} + \frac{B}{(1+x)}$$

$$\equiv \frac{A(1+x)+B(2-x)}{(2-x)(1+x)}$$

Equating the numerators $A(1+x)+B(2-x) \equiv 6$.

Hence $A+2B = 6$ and $A-B = 0 \Rightarrow A = B = 2$.

$$\therefore \quad \text{Area} = \int_0^1 3\,dx - \int_0^1 \frac{2}{(2-x)}\,dx - \int_0^1 \frac{2}{(1+x)}\,dx$$

$$= [3x+2\ln|2-x|-2\ln|1+x|]_0^1$$

$$= (3+2\ln 1-2\ln 2)-(0+2\ln 2-2\ln 1)$$

$= 3-4\ln 2 = 0.227$ to three decimal places.

(b) The area bounded by the curve, the x-axis and the lines $x = 4$ and $x = 5$ is given by

$$\int_4^5 \frac{2}{(2-x)} + \frac{2}{(1+x)}\,dx = [-2\ln|2-x| + 2\ln|1+x|]_4^5$$

$$= \left[2\ln\left|\frac{1+x}{2-x}\right|\right]_4^5$$

$$= 2\ln|-\tfrac{6}{3}| - 2\ln|-\tfrac{5}{2}|$$

$$= 2\ln\tfrac{4}{5} = -0.446 \text{ to three decimal places}$$

Chapter 5
Exponential and Logarithmic Functions

Multiple choice questions (Type A) Select the correct answer

Example 1

$\log_a b =$

A $\dfrac{\log_b c}{\log_a c}$ **B** $\dfrac{\log_a c}{\log_c b}$ **C** $\dfrac{\log_c a}{\log_c b}$ **D** $\dfrac{\log_c b}{\log_a c}$ **E** $\dfrac{\log_c b}{\log_c a}$

Let $L = \log_a b \Rightarrow a^L = b$ using logarithm property

$\Rightarrow \log_c a^L = \log_c b$ taking logs to base c

$\Rightarrow L \log_c a = \log_c b$ using logarithm property

$\Rightarrow L = \dfrac{\log_c b}{\log_c a}$ ANSWER E

Example 2

$\dfrac{d}{dx}(\ln \tan x) =$

A $\dfrac{1}{\tan x}$ **B** $\dfrac{1}{\sin x} - \dfrac{1}{\cos x}$ **C** $\dfrac{\sec^2 x}{\tan x}$ **D** $-\dfrac{\sec x}{\tan x}$ **E** $\dfrac{\cot x}{\tan x}$

Put $u = \tan x$, so $y = \ln \tan x = \ln u$ and $\dfrac{dy}{du} = \dfrac{1}{u}$ and $\dfrac{du}{dx} = \sec^2 x$.

$$\frac{dy}{dx} = \frac{dy}{du} \times \frac{du}{dx} = \frac{1}{u} \times \sec^2 x = \frac{\sec^2 x}{\tan x}$$ ANSWER C

Alternatively, $y = \ln \tan x = \ln(\sin x/\cos x) = \ln \sin x - \ln \cos x$

$$\frac{dy}{dx} = \frac{\cos x}{\sin x} + \frac{\sin x}{\cos x} = \frac{\cos^2 x + \sin^2 x}{\sin x \cos x} = \frac{1}{\sin x \cos x} = \frac{\sec^2 x}{\tan x}$$

Example 3

$\dfrac{d}{dx}(\log_{10} x) =$

A $\dfrac{1}{x}$ **B** $\dfrac{1}{x}\log_{10} e$ **C** $\dfrac{1}{x}\log_e 10$ **D** $\dfrac{1}{10x}$ **E** $\dfrac{10}{x}$

Use the change of base in Question 1 if you cannot remember the result.

$$y = \log_{10}x = \frac{\log_e x}{\log_e 10} = \frac{\ln x}{\ln 10} \Rightarrow \frac{dy}{dx} = \frac{1}{x \ln 10}$$

This does not appear as one of the alternative answers.

However $\ln 10 = \log_e 10 = \dfrac{\log_{10}10}{\log_{10}e} = \dfrac{1}{\log_{10}e}$ so $\dfrac{1}{x \ln 10} = \dfrac{1}{x}\log_{10}e$

$$\frac{dy}{dx} = \frac{1}{x} \times \frac{1}{\log_e 10} = \frac{1}{x}\log_{10}e$$

ANSWER B

Example 4
The coefficient of x^3 in the expansion of e^{2x} is

A $\frac{1}{6}$ **B** $-\frac{1}{6}$ **C** $\frac{4}{3}$ **D** $-\frac{4}{3}$ **E** 8

$$e^{2x} = 1 + 2x + \frac{(2x)^2}{2!} + \frac{(2x)^3}{3!} + \dots$$

The coefficient of x^3 is $\dfrac{2^3}{3!} = \dfrac{8}{6} = \dfrac{4}{3}$ ANSWER C

Example 5
In the expansion of $\ln(1-2x)$ the first two non-zero terms are

A $-2x+2x^2$ **B** $-2x-2x^2$ **C** $2x-4x^2$ **D** $2x+4x^2$
E $-2x-x^2$

$\ln(1+x) = x - \frac{1}{2}x^2 + \frac{1}{3}x^3 - \dots$
$\Rightarrow \ln(1-2x) = -2x - \frac{1}{2}\cdot 4x^2 - \dots = -2x - 2x^2$ ANSWER B

Example 6
For the relation $y = ax^n$ a straight line is obtained by plotting

A y against $\log x$ **B** $\log y$ against x **C** y against x
D $\log y$ against $\log x$ **E** none of these

$$y = ax^n \Rightarrow \log y = \log ax^n = \log a + \log x^n = \log a + n \log x$$

Plot $\log y$ on the y-axis and $\log x$ on the x-axis to give $y = nx + \log a$, which is a straight line, of gradient n and y intercept $\log a$. ANSWER D

From the graph the numerical values of n and a are calculated.

Example 7
Which of the following functions is even?

A e^x **B** e^{-x} **C** $\ln x$ **D** e^x+e^{-x} **E** e^x-e^{-x}

An even function satisfies $f(x)=f(-x)$ and is symmetrical about the y-axis.
$e^x \neq e^{-x}$ which rules out answers A and B; $\ln(-x)$ is not defined.

$$f(x) = e^x+e^{-x} \Rightarrow f(-x) = e^{-x}+e^x = f(x)$$

ANSWER D

The function in answer D is $2\cosh x$, which is even.
The function in answer E is $2\sinh x$, which is odd, i.e., $f(-x) = -f(x)$.

Example 8
The sum to n terms of the series $\ln e+\ln e^2+\ldots+\ln e^n$ is

A $\dfrac{e^n-1}{e-1}$ **B** $\dfrac{1}{e-1}$ **C** $\dfrac{1}{\ln e-1}$ **D** $\dfrac{\ln e^n-1}{\ln e-1}$ **E** $\dfrac{n(n+1)}{2}$

$$\ln e+\ln e^2+\ln e^3+\ldots+\ln e^n = 1+2+3+\ldots+n = \tfrac{1}{2}n(n+1)$$

ANSWER E

Example 9
$\log_2 32 \div \log_2 4 =$

A 4 **B** $2\frac{1}{2}$ **C** 28 **D** 3 **E** 2^3

$$\log_2 32 \div \log_2 4 = \log_2 2^5 \div \log_2 2^2 = 5\log_2 2 \div 2\log_2 2 = 5 \div 2 = 2\tfrac{1}{2}$$

ANSWER B

Multiple choice questions (Type B) Answer according to the table

A	B	C	D	E
1, 2, 3 correct	1, 3 only	2, 3 only	2 only	3 only

Example 10
$\dfrac{d}{dx}(\ln x^3) =$

1 $\dfrac{1}{x^3}$ **2** $\dfrac{3}{x}$ **3** $\dfrac{3x^2}{x^3}$

$$\ln x^3 = 3\ln x \Rightarrow \frac{d}{dx}(\ln x^3) = \frac{3}{x} = \frac{3x^2}{x^3}$$

Answers 2 and 3 are correct. ANSWER C

Example 11

$$I = \int_2^p \frac{1}{x}\,dx = k - \ln 2$$

1 $p = 4 \Rightarrow I = \ln 2$ **2** $p = 1 \Rightarrow k = 0$ **3** $p = \mathrm{e} \Rightarrow k = 1$

$$p = 4 \Rightarrow I = \int_2^p \frac{1}{x}\,dx = \ln 4 - \ln 2 = \ln 2$$

Statement 1 is correct

$p = 1 \Rightarrow k = \ln 1 = 0$ Statement 2 is correct

$p = \mathrm{e} \Rightarrow k = \ln \mathrm{e} = 1$ Statement 3 is correct

ANSWER A

Example 12

$$\frac{d}{dx}\{\ln(\sec x + \tan x)\} =$$

1 $\sec x$ **2** $\dfrac{1}{\sec x + \tan x}$ **3** $\dfrac{\sec x \tan x + \sec^2 x}{\sec x + \tan x}$

$$\frac{d}{dx}\ln(\sec x + \tan x) = \frac{\sec x \tan x + \sec^2 x}{\sec x + \tan x}$$

statement 3 is correct

$$= \frac{\sec x(\tan x + \sec x)}{\sec x + \tan x}$$

$= \sec x$ statement 1 is correct

statement 2 is wrong

ANSWER B

Example 13

For the graph of $x^2\mathrm{e}^{-x}$

1 $x = 2$ gives a maximum point **2** $y \to 0$ as $x \to -\infty$
3 $y \to 0$ as $x \to +\infty$

$$y = x^2e^{-x} \Rightarrow \frac{dy}{dx} = 2xe^{-x} - x^2e^{-x} = x(2-x)e^{-x} = 0 \quad \text{when } x = 0, 2.$$

$$\frac{d^2y}{dx^2} = 2e^{-x} - 4xe^{-x} + x^2e^{-x} = (x^2 - 4x + 2)e^{-x} < 0 \quad \text{when} \quad x = 2$$

$\Rightarrow$ maximum. Statement 1 is correct.

As $x \to \infty$, $e^x > x^2 \Rightarrow \dfrac{x^2}{e^x} \to 0$, i.e., $x^2e^{-x} \to 0$. Statement 3 is correct.

As $x \to -\infty$, e^{-x} and x^2 are both large and positive.
Statement 2 is wrong. ANSWER B

As $x \to \infty$, e^x gets larger more quickly than x^2

$$\Rightarrow \frac{x^2}{e^x} = 0, \text{ i.e., } x^2e^{-x} = 0$$

Statement 3 is correct.

General questions

Example 14
Solve the equations (a) $e^{\ln x} = 5$ and (b) $\ln e^x = 6$.

The usual way to solve equations involving powers is to take logarithms in order to eliminate the powers.

(a) $\ln(e^{\ln x}) = \ln 5$
$\ln x \cdot \ln e = \ln 5$
$\ln x = \ln 5$ since $\ln e = 1$
$x = 5$

(b) $\ln e^x = 6$
$x \ln e = 6$
$x = 6$

In fact e^x and $\ln x$ are inverse functions, so when applied successively they produce the identity function, i.e.,

$$e^{\ln x} = x \quad \text{and} \quad \ln e^x = x$$

If you can remember these, the equations are trivial.

Example 15
The product rule for differentiation is $\dfrac{d}{dx}(uv) = u\dfrac{dv}{dx} + v\dfrac{du}{dx}$. Derive the product rule for differentiating a product of 3 functions uvw by using logarithms.

$$f = uvw \Rightarrow \ln f = \ln(uvw) = \ln u + \ln v + \ln w$$

$$\Rightarrow \frac{1}{f}\frac{df}{dx} = \frac{1}{u}\frac{du}{dx} + \frac{1}{v}\frac{dv}{dx} + \frac{1}{w}\frac{dw}{dx}$$

$$\Rightarrow \frac{df}{dx} = \frac{f}{u}\frac{du}{dx} + \frac{f}{v}\frac{dv}{dx} + \frac{f}{w}\frac{dw}{dx}$$

$$\Rightarrow \frac{d}{dx}(uvw) = vw\frac{du}{dx} + uw\frac{dv}{dx} + uv\frac{dw}{dx}$$

Example 16

Find $\int \frac{2x}{x^2-1}\,dx$ (a) directly and (b) by partial fractions.

(a) You must recognize that the numerator is the derivative of the denominator and since $\int \frac{f'(x)}{f(x)}\,dx = \ln f(x)$.

$$\int \frac{2x}{x^2-1}\,dx = \ln(x^2-1) + c$$

(b) $\frac{2x}{x^2-1} = \frac{2x}{(x-1)(x+1)} = \frac{a}{x-1} + \frac{b}{x+1} \Rightarrow 2x = a(x+1) + b(x-1)$

By covering up or using $x = 1$ and -1, $a = 1$ and $b = 1$

$$\int \frac{2x}{x-1}\,dx = \int\left(\frac{1}{x-1} + \frac{1}{x+1}\right)dx = \ln(x-1) + \ln(x+1) + c$$
$$= \ln(x^2-1) + c$$

Example 17

Obtain the expansion for $\ln(1+x)$ in ascending powers of x and deduce the expansion for $\ln(1-x)$. Use these two to work out $\ln(1-x^2)$ and check this by comparing it with $\ln(1-x)$ with x^2 replacing x.

$$f(x) = f(0) + xf'(0) + \frac{x^2}{2!}f''(0) + \ldots$$

$$f(x) = \ln(1+x) \Rightarrow f'(x) = \frac{1}{1+x},\ f''(x) = \frac{-1}{(1+x)^2},$$

$$f'''(x) = \frac{2}{(1+x)^3}$$

$$f''''(x) = \frac{-3!}{(1+x)^4}, \; f^r(x) = \frac{(-1)^{r-1}(r-1)!}{(1+x)^r}$$

$x = 0 \Rightarrow f(0) = 0, f'(0) = 1, f''(0) = -1, f'''(0) = 2, f''''(0) = -3!$, etc.

$$\ln(1+x) = x - \frac{x^2}{2} + \frac{2x^3}{3!} - \frac{3!x^4}{4!} + \frac{4!x^5}{5!}$$

$$= x - \frac{x^2}{2} + \frac{x^3}{3} - \frac{x^4}{4} + \frac{x^5}{5} - \ldots + \frac{(-1)^{r-1}x^r}{r} \qquad (1)$$

Substituting $-x$ for x gives

$$\ln(1-x) = -x - \tfrac{1}{2}x^2 - \tfrac{1}{3}x^3 - \tfrac{1}{4}x^4 - \ldots \qquad (2)$$

Adding (1) and (2) gives $\ln(1+x) + \ln(1-x) = \ln(1+x)(1-x) = \ln(1-x^2)$. Equations (1) and (2) give $\ln(1-x) = -x^2 - \tfrac{1}{2}x^4 - \tfrac{1}{3}x^6 - \tfrac{1}{4}x^8$ which is just the same as the series obtained by substituting x^2 for x in (2).

Example 18

Find $\frac{d}{dx}\{\tfrac{1}{4}(e^x + e^{-x})^2 - \tfrac{1}{4}(e^x - e^{-x})^2\}$

Some students may recognize this as

$$\frac{d}{dx}(\cosh^2 x - \sinh^2 x) = \frac{d}{dx}(1) = 0$$

Otherwise the difference of two squares will give

$$\frac{d}{dx}(\tfrac{1}{4}(e^x + e^{-x})^2 - \tfrac{1}{4}(e^x - e^{-x})^2) = \frac{d}{dx}\left(\frac{2e^x}{2} \times \frac{2e^{-x}}{2}\right) = \frac{d}{dx}(1) = 0$$

Example 19

For $x > 0$, find the minimum point on the curve $y = x^x$.

$$\ln y = \ln x^x = x \ln x \Rightarrow \frac{1}{y}\frac{dy}{dx} = \ln x + x \cdot \frac{1}{x} = 1 + \ln x$$

$$\frac{dy}{dx} = y(1 + \ln x) = x^x(1 + \ln x) = 0 \Rightarrow \ln x = -1 \Rightarrow x = e^{-1} = \frac{1}{e} = 0.37$$

$$x = e^{-1} \Rightarrow y = \left(\frac{1}{e}\right)^{1/e} = 0.69$$

$$\frac{d^2y}{dx^2} = x^x(1+\ln x)^2 + x^x \cdot \frac{1}{x} > 0 \text{ when } x = 0.37$$

Minimum point is (0.37,0.69).

Example 20
(a) $\int xe^x\,dx =$ (b) $\int x\ln x\,dx =$ (c) $\int \ln x\,dx =$

(a) This involves integrating a product $\int fg' = fg - \int f'g\,dx$

Let $f = x, g' = e^x \Rightarrow f' = 1$ and $g = \int e^x\,dx = e^x$

$$\int xe^x\,dx = xe^x - \int 1 \times e^x\,dx = xe^x - e^x + c$$

When integrating products (integration by parts) one function is differentiated and one integrated. The choice of which function to integrate and which to differentiate depends on whether the second integral ($f'g\,dx$ above) is made easier. In this example f is chosen to be x so that f' is 1 and the second integral is no longer a product.

(b) In this example, you have no choice for f or g' since you cannot integrate $\ln x$ so $\ln x$ must differentiate, i.e., $f = \ln x$.

$$f = \ln x \text{ and } g' = x \Rightarrow f' = \frac{1}{x} \text{ and } g = \tfrac{1}{2}x^2$$

$$\int x\ln x\,dx = \tfrac{1}{2}x^2 \times \ln x - \int \tfrac{1}{2}x^2 \times \frac{1}{x}\,dx = \tfrac{1}{2}x^2 \ln x - \int \tfrac{1}{2}x\,dx$$
$$= \tfrac{1}{2}x^2 \ln x - \tfrac{1}{4}x^2 + c$$

(c) In this example a product has to be created since $\ln x$ cannot be integrated directly.

$$f = \ln x \text{ and } g' = 1 \Rightarrow f' = \frac{1}{x} \text{ and } g = \int 1\,dx = x$$

$$1 \times \ln x\,dx = x\ln x - x\frac{1}{x}\,dx = x\ln x - x + c$$

Chapter 6
Coordinate Geometry

This is the study of the properties of lines, curves and planes using the methods of algebra and calculus. The topics included in this branch of the subject differ considerably according to the syllabus being followed. It is possible that only the straight line and the circle will be needed although candidates may be expected to have knowledge of the equations of the conics. Parameters are usually included and students must be able to deal with equations of tangents, normals and chords.

Multiple choice questions (Type A) Select the correct answer

> **Example 1**
> The equation of the straight line in Figure 13 is
>
> **A** $2y-3x=6$ **B** $3y+2x=9$ **C** $2y+3x=9$
> **D** $2y+3x=6$ **E** $3y-2x=9$

The gradient of the line is

$$\frac{3-0}{0-2}=-\frac{3}{2}$$

The equation of a line of gradient m passing through the point (x_1, y_1) is

$$y-y_1=m(x-x_1)$$

This line passes through $(0, 3)$ so its equation is

$$y-3=-\tfrac{3}{2}(x-0)$$
$$\Rightarrow 2y+3x=6$$

ANSWER D

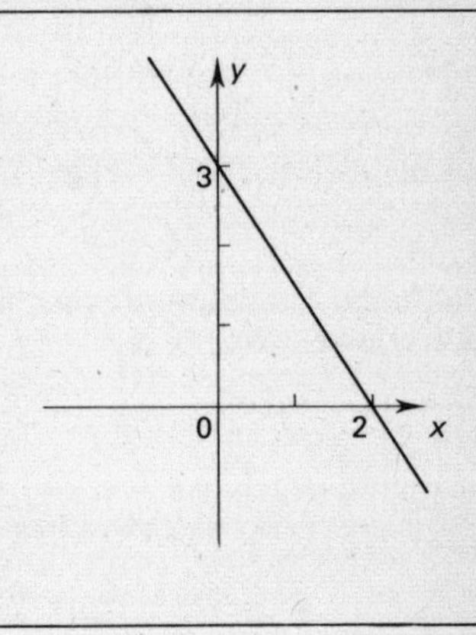

Figure 13

> **Example 2**
> The coordinates of the point C which divides the line segment joining the points $A(-3, -4)$ and $B(-1, 6)$ internally in the ratio of $3:1$ are
>
> **A** $(-2\frac{1}{2}, -1\frac{1}{2})$ **B** $(-1\frac{1}{2}, 3\frac{1}{2})$ **C** $(-2\frac{2}{3}, -1\frac{1}{3})$
> **D** $(-1\frac{2}{3}, 2\frac{2}{3})$ **E** $(-1\frac{1}{2}, -2\frac{1}{2})$

From Figure 14 it can be seen that triangles ADC and AFB are similar.

Hence $$\frac{AC}{AB} = \frac{AD}{AF} = \frac{DC}{FB} = \tfrac{3}{4}$$

$\therefore \quad AD = \tfrac{3}{4}AF = \tfrac{3}{4}\{-1-(-3)\} = \tfrac{3}{4}\times 2 \ = 1\tfrac{1}{2}$

and $DC = \tfrac{3}{4}FB = \tfrac{3}{4}\{6-(-4)\} \quad = \tfrac{3}{4}\times 10 = 7\tfrac{1}{2}$

Therefore, the coordinates of C are

$$((-3+1\tfrac{1}{2}),(-4+7\tfrac{1}{2})) = (-1\tfrac{1}{2}, 3\tfrac{1}{2})$$ ANSWER B

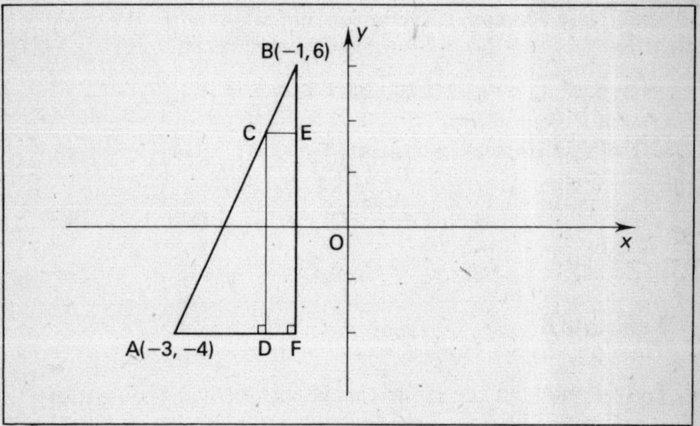

Figure 14

Example 3

The equation of the straight line perpendicular to the line $3x+4y-2=0$ at the point $(6, -4)$ is

A $3y-4x+36=0$ **B** $3y+4x-12=0$ **C** $3y-4x-34=0$
D $3y+4x-2=0$ **E** $3y-4x+12=0$

The product of the gradients of two perpendicular lines is -1. The gradient of the line $3x+4y-2=0$ is $-\tfrac{3}{4}$ and thus the gradient of the perpendicular is $\tfrac{4}{3}$.

Since this line passes through the point $(6, -4)$ its equation is

$$y-(-4) = \tfrac{4}{3}(x-6) \Rightarrow 3y+12 = 4x-24 \Rightarrow 3y-4x+36=0$$

ANSWER A

Example 4
The centre of the circle $2x^2+2y^2-3x+5y-8=0$ is

A $(-3,5)$ **B** $(-\frac{3}{2},\frac{5}{2})$ **C** $(-\frac{3}{4},\frac{5}{4})$ **D** $(\frac{3}{2},-\frac{5}{2})$ **E** $(\frac{3}{4},-\frac{5}{4})$

The given equation can be written as $x^2+y^2-\frac{3}{2}x+\frac{5}{2}y-4=0$.
This gives $(x-\frac{3}{4})^2+(y+\frac{5}{4})^2=4+\frac{34}{16}=\frac{49}{8}$.

This is the equation of a circle centre $(\frac{3}{4},-\frac{5}{4})$, radius $\dfrac{7\sqrt{2}}{4}$.

ANSWER E

Alternatively, compare the given equation with the standard equation $x^2+y^2+2gx+2fy+c=0$ where the centre is $(-g,-f)$.

Clearly, $g=-\frac{3}{4}$ and $f=\frac{5}{4}$ and the centre is $(\frac{3}{4},-\frac{5}{4})$.

Multiple choice questions (Type B) Answer according to the table

A	B	C	D	E
1, 2, 3 correct	1, 3 only	2, 3 only	2 only	3 only

Example 5
The equation of a circle is $x^2+y^2=8$.

1 Any point on the circle can be represented in terms of a parameter θ by $(2\sqrt{2}\cos\theta, 2\sqrt{2}\sin\theta)$.
2 The equation of the tangent at $(2,2)$ is $x+y=4$.
3 The line $y=-x-4$ is a tangent to the circle.

If θ is the parameter then $x=2\sqrt{2}\cos\theta$ and $y=2\sqrt{2}\sin\theta$.

Thus $\cos\theta=\dfrac{x}{2\sqrt{2}}$ and $\sin\theta=\dfrac{y}{2\sqrt{2}}$.

Squaring and adding $\dfrac{x^2}{8}+\dfrac{y^2}{8}=\cos^2\theta+\sin^2\theta=1$

$\therefore$ $x^2+y^2=8$ and the point $(2\sqrt{2}\cos\theta, 2\sqrt{2}\sin\theta)$ satisfies the equation for all values of θ.
Statement 1 is correct.

If $x^2+y^2=8$ then, differentiating implicitly with respect to x, we have $2x+2y\dfrac{dy}{dx}=0 \Rightarrow \dfrac{dy}{dx}=-\dfrac{x}{y}$.

At $(2, 2)$ the gradient of the tangent is therefore $-\frac{2}{2} = -1$.

Hence the equation of the tangent is

$$y-2 = -1(x-2) \Rightarrow x+y = 4$$

Statement 2 is correct.

Consider the intersection of the line $y = -x-4$ and the circle. Solving simultaneously

$$x^2+(-x-4)^2 = 8 \Rightarrow x^2+x^2+8x+16 = 8$$

$$\therefore\ 2x^2+8x+8 = 0 \Rightarrow x^2+4x+4 = 0 \text{ or } (x+2)^2 = 0.$$

Since this produces a repeated root, the two cutting points are coincident and the line is a tangent to the circle.

Statement 3 is correct.
Statements 1, 2 and 3 are all true. ANSWER A

General questions

These questions illustrate some of the points that may be raised as a short problem in itself or as a longer question involving different parts.

Example 6

Find the coordinates of the points of intersection of the circles

$$x^2+y^2+2x+4y-3 = 0 \text{ and } x^2+y^2+3x+5y = 0$$

If $x^2+y^2+2x+4y-3 = 0$ and $x^2+y^2+3x+5y = 0$ then, subtracting, we have

$$-x-y-3 = 0 \Rightarrow y = -3-x \qquad (1)$$

Substituting in the first equation

$$x^2+(-3-x)^2+2x+4(-3-x)-3 = 0$$

$$\therefore\ 2x^2+4x-6 = 0 \Rightarrow x^2+2x-3 = 0$$

$$\therefore\ (x+3)(x-1) = 0 \Rightarrow x = -3 \text{ or } x = 1$$

If $x = -3$, $y = 0$ and if $x = 1$, $y = 4$ (using (1)).

The coordinates of the points of intersection are $(-3, 0)$, $(1, -4)$.

Note that the line $x+y = -3$ is called the radical axis of the two circles. In general, if $S_1 = 0$ and $S_2 = 0$ are the equations of two intersecting circles then $S_1+\lambda S_2 = 0$ represents a circle through the points of intersection for all $\lambda \neq -1$. When $\lambda = -1$, $S_1-S_2 = 0$ is a straight line.

Example 7

Find the equation of the circle through the points $A(1, 1)$, $B(2, 0)$ and $C(-1, -3)$.

Let the equation of the circle be $x^2+y^2+2ax+2by+c = 0$.

$A(1, 1)$ lies on the circle	$2a+2b+c+2 = 0$	(1)
$B(2, 0)$ lies on the circle	$4a+c+4 = 0$	(2)
$C(-1, -3)$ lies on the circle	$-2a-6b+c+10 = 0$	(3)
Eliminating c from (1) and (2)	$2a-2b+2 = 0$	(4)
Eliminating c from (2) and (3)	$-6a-6b+6 = 0$	(5)

Multiplying (4) by 3 and subtracting (5) gives $12a = 0 \Rightarrow a = 0$.
Hence $b = 1$ and $c = -4$.
The equation of the circle is $x^2+y^2+2y-4 = 0$.

Example 8
Show that if the tangents to the curve $y^2 = 4x$ at the points $P(p^2, 2p)$ and $Q(q^2, 2q)$ contain an angle of $\frac{3}{4}\pi$ then $q = \dfrac{(p-1)}{(p+1)}$.

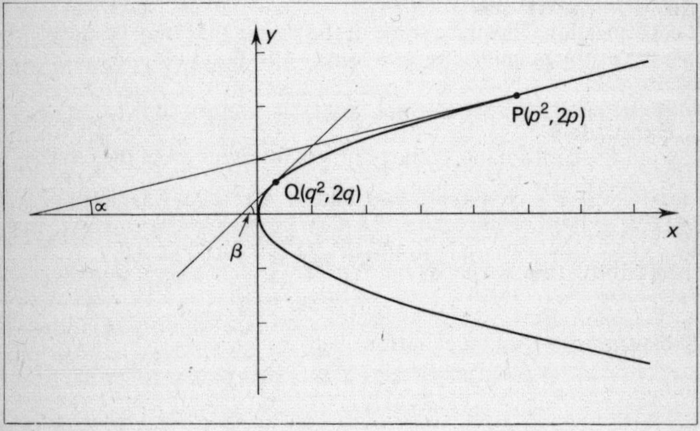

Figure 15

Since $y^2 = 4x$, differentiating implicitly with respect to x

$$2y\frac{dy}{dx} = 4 \Rightarrow \frac{dy}{dx} = \frac{2}{y}$$

Hence the gradients of the tangents at P and Q are $\frac{1}{p}$ and $\frac{1}{q}$ respectively.

If α and β represent the angles that the tangents make with the x-axis then $\tan\alpha = \frac{1}{p}$ and $\tan\beta = \frac{1}{q}$.

But $\beta-\alpha = 45°$ $\therefore$ $\tan(\beta-\alpha) = \tan 45°$

$$\therefore \quad \frac{\tan\beta - \tan\alpha}{1+\tan\beta\tan\alpha} = \frac{\frac{1}{q}-\frac{1}{p}}{1+\frac{1}{q}\cdot\frac{1}{p}} = \frac{p-q}{pq+1} = 1$$

Simplifying we have

$$pq+1 = p-q \Rightarrow q(p+1) = p-1 \text{ or } q = \frac{p-1}{p+1}.$$

Example 9
Show that if $P(ap^2, 2ap)$ and $Q(aq^2, 2aq)$ are two points on the parabola $y^2 = 4ax$ such that PQ passes through the point $(a, 0)$ then the area of the triangle OPQ is $a^2(p-q)$ where O is the vertex of the parabola.

The diagram is shown in Figure 16.

The gradient of the chord $PQ = \frac{2ap-2aq}{ap^2-aq^2} = \frac{2a(p-q)}{a(p^2-q^2)} = \frac{2}{p+q}$.

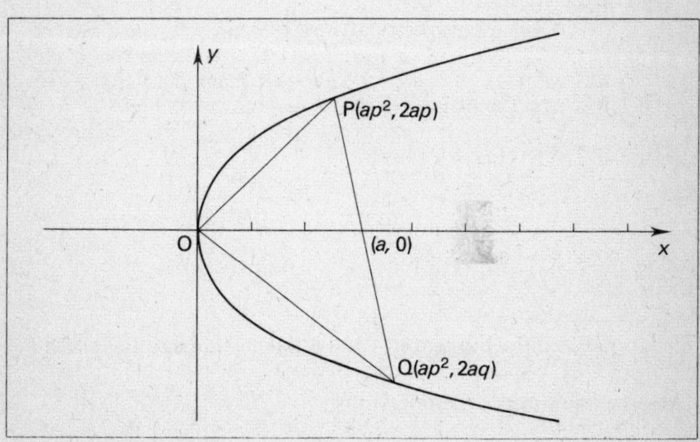

Figure 16

The equation of the chord PQ is

$$y-2ap = \frac{2}{p+q}(x-ap^2) \text{ or } 2x-(p+q)y+2apq = 0$$

Since the point $(a, 0)$ lies on the line,

$$\therefore \quad 2a+2apq = 0 \Rightarrow pq = -1$$

$\therefore$ the equation of the chord is $2x-(p+q)y-2a = 0$.

The distance of the origin from this line is given by

$$ON = \frac{2a}{\sqrt{4+(p+q)^2}}$$

The length of $PQ = \sqrt{\{2a(p-q)\}^2+(ap^2-aq^2)^2}$

$$= a(p-q)\sqrt{4+(p+q)^2}$$

$\therefore$ area of triangle $OPQ = \frac{1}{2}PQ \times ON$

$$= \tfrac{1}{2}a(p-q)\sqrt{4+(p+q)^2} \times \frac{2a}{\sqrt{4+(p+q)^2}}$$

$$= a^2(p-q)$$

Sometimes examination questions may ask for the derivation of the standard equations of tangents and normals to curves as the first part of a problem.

Example 10
A curve is defined parametrically by $x = at^2, y = 2at$. Find the equation of the tangent at $P(at^2, 2at)$. If this meets the x-axis at R and the foot of the perpendicular from P to the x-axis is Q show that the origin O is the mid-point of RQ.

Since $x = at^2 \Rightarrow \dfrac{dx}{dt} = 2at$ and $y = 2at \Rightarrow \dfrac{dy}{dt} = 2a$.

Hence $\dfrac{dy}{dx} = \dfrac{dy}{dt} \times \dfrac{dt}{dx} = \dfrac{2a}{2at} = \dfrac{1}{t}$.

The gradient of the tangent is $\dfrac{1}{t}$ and it passes through the point P.

Thus the equation of the tangent is

$$y-2at = \frac{1}{t}(x-at^2) \qquad \text{or} \qquad ty = x+at^2$$

The coordinates of R are $(-at^2, 0)$.
The coordinates of Q are $(at^2, 0)$.
The mid-point of RQ is $(\frac{1}{2}(at^2-at^2), 0) = (0, 0)$.
Thus the origin is the mid-point of RQ.

Example 11
Find the equation of the normal to the curve given by $x = 3\cos\theta, y = 2\sin\theta$ at the point $P(\frac{3}{2}, \sqrt{3})$. Hence find the coordinates of the points A and B in which this normal meets the axes.

Differentiating parametrically

$$\frac{dy}{d\theta} = 2\cos\theta \text{ and } \frac{dx}{d\theta} = -3\sin\theta \Rightarrow \frac{d\theta}{dx} = -\frac{1}{3\sin\theta}$$

Hence
$$\frac{dy}{dx} = \frac{dy}{d\theta}\cdot\frac{d\theta}{dx} = -\frac{2\cos\theta}{3\sin\theta}$$

The gradient of the tangent at $P(\frac{3}{2}, \sqrt{3})$ where $\theta = \frac{1}{3}\pi$ is $-\dfrac{2}{3\sqrt{3}}$.

The gradient of the normal at P is $\dfrac{3\sqrt{3}}{2}$.
The equation of the normal is

$$y - \sqrt{3} = \frac{3\sqrt{3}}{2}(x - \tfrac{3}{2}) \text{ or } 4\sqrt{3}y = 18x - 15$$

If this line cuts the x-axis at A and the y-axis at B then A is the point $(\frac{5}{6}, 0)$ and B is the point $\left(0, -\dfrac{5\sqrt{3}}{4}\right)$.

Example 12
Find the equation of the normal to the curve $x = ct, y = \dfrac{c}{t}$ at the point P whose parameter is p. If this normal meets the curve again at Q, find the coordinates of Q and the locus of the mid-point of PQ as P varies.

Eliminating t we have $xy = c^2 \Rightarrow y = \dfrac{c^2}{x} \Rightarrow \dfrac{dy}{dx} = -\dfrac{c^2}{x^2}$.

Thus the gradient of the tangent at $P\left(cp, \dfrac{c}{p}\right)$ is $-\dfrac{1}{p^2}$.
The gradient of the normal at P is p^2.
The equation of the normal at P is

$$y - \frac{c}{p} = p^2(x - cp) \text{ or } p^3x - py + c(1 - p^4) = 0$$

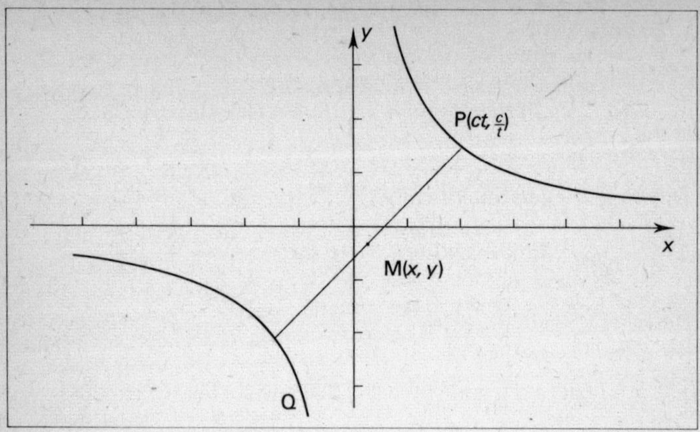

Figure 17

This line cuts the curve $xy = c^2$ again at Q and hence, solving the two equations simultaneously

$$p^3x - \frac{pc^2}{x} + c(1-p^4) = 0 \Rightarrow p^3x^2 + c(1-p^4)x - c^2p = 0$$

Since $x = cp$ is a solution $(x-cp)(p^3x+c) = 0$.

Hence $x = -\dfrac{c}{p^3}$ is the x coordinate of $Q \Rightarrow Q$ is at $\left(-\dfrac{c}{p^3}, -cp^3\right)$.

Let $M(x, y)$ be the mid-point of PQ.

Thus $\quad x = \dfrac{1}{2}\left\{cp + \left(-\dfrac{c}{p^3}\right)\right\}$ and $y = \dfrac{1}{2}\left\{\dfrac{c}{p} + (-cp^3)\right\}$

or $\quad 2xp^3 = c(p^4-1) \quad (1)$ and $2yp = c(1-p^4) \quad (2)$

Combining (1) and (2) $2xp^3 = -2yp \Rightarrow p^2 = -\dfrac{y}{x}$

Substituting in (2) after squaring

$$4y^2p^2 = c^2(1-p^4)^2 \Rightarrow 4y^2\left(-\frac{y}{x}\right) = c^2\left(1 - \frac{y^2}{x^2}\right)^2$$

The equation of the locus is $4x^3y^3 = c^2(x^2-y^2)$.

Chapter 7
Matrices and Determinants

Multiple choice questions (Type A) Select the correct answer

Example 1
Which matrix represents a rotation?

A $\begin{pmatrix} -1 & 0 \\ 0 & -1 \end{pmatrix}$ **B** $\begin{pmatrix} 0 & -1 \\ -1 & 0 \end{pmatrix}$ **C** $\begin{pmatrix} -1 & 0 \\ 0 & 1 \end{pmatrix}$ **D** $\begin{pmatrix} 1 & 0 \\ 0 & -1 \end{pmatrix}$

E $\begin{pmatrix} 0 & 1 \\ 1 & 0 \end{pmatrix}$

The simplest way to answer the question is to find the matrix whose determinant is equal to 1 as this is a necessary condition for a rotation. The determinant gives the area scale factor. **A** is the only matrix whose determinant is equal to 1. The others all have determinants equal to -1 and represent reflexions. ANSWER A

In other questions you may have to find the image of the unit square. Remember that the first matrix column gives the image of the point (1,0) and the second column gives the image of (0,1). For matrix **A**, $(1,0)\rightarrow(-1,0)$ and $(0,1)\rightarrow(0,-1)$. This is a rotation of 180° about the origin.

Example 2
The inverse of the matrix $\begin{pmatrix} 3 & 1 \\ 4 & 2 \end{pmatrix}$ is

A $\begin{pmatrix} 3 & -1 \\ -4 & 2 \end{pmatrix}$ **B** $\begin{pmatrix} -3 & 1 \\ 4 & -2 \end{pmatrix}$ **C** $\begin{pmatrix} 2 & -1 \\ -4 & 3 \end{pmatrix}$

D $\begin{pmatrix} -2 & 1 \\ 4 & -3 \end{pmatrix}$ **E** $\begin{pmatrix} 1 & -\frac{1}{2} \\ -2 & 1\frac{1}{2} \end{pmatrix}$

The inverse of $\begin{pmatrix} a & b \\ c & d \end{pmatrix}$ is $\frac{1}{\Delta}\begin{pmatrix} d & -b \\ -c & a \end{pmatrix} = \begin{pmatrix} d/\Delta & -b/\Delta \\ -c/\Delta & a/\Delta \end{pmatrix}$ where $\triangle = ad - bc$.

The inverse of $\begin{pmatrix} 3 & 1 \\ 4 & 2 \end{pmatrix}$ is $\frac{1}{2}\begin{pmatrix} 2 & -1 \\ -2 & 3 \end{pmatrix} = \begin{pmatrix} 1 & -\frac{1}{2} \\ -2 & 1\frac{1}{2} \end{pmatrix}$

ANSWER E

Example 3
Which matrix represents a translation?

A $\begin{pmatrix} 2 & 0 \\ 3 & 0 \end{pmatrix}$ **B** $\begin{pmatrix} 2 & 3 \\ 0 & 0 \end{pmatrix}$ **C** $\begin{pmatrix} 2 & 0 \\ 0 & 3 \end{pmatrix}$ **D** $\begin{pmatrix} 0 & 2 \\ 3 & 0 \end{pmatrix}$

E none of these

For a transformation represented by the matrix $\begin{pmatrix} a & b \\ c & d \end{pmatrix}$, (x,y) moves to (X, Y) where $\begin{pmatrix} X \\ Y \end{pmatrix} = \begin{pmatrix} a & b \\ c & d \end{pmatrix}\begin{pmatrix} x \\ y \end{pmatrix}$. So $X = ax+by$ and $Y = cx+dy$.

The translation $\begin{pmatrix} 2 \\ 3 \end{pmatrix}$ or $2\mathbf{i}+3\mathbf{j}$ which takes (x,y) to $(x+2, y+3)$, cannot be represented by premultiplying by a 2×2 matrix. Another reason is that any 2×2 matrix transformation keeps the origin fixed, whereas it would need to move in a translation.

ANSWER E

Example 4
Which matrix represents, in three dimensions, a rotation of 180° about the x-axis.

A $\begin{pmatrix} 1 & 0 & 0 \\ 0 & -1 & 0 \\ 0 & 0 & 1 \end{pmatrix}$ **B** $\begin{pmatrix} -1 & 0 & 0 \\ 0 & 1 & 0 \\ 0 & 0 & -1 \end{pmatrix}$ **C** $\begin{pmatrix} 1 & 0 & 0 \\ 0 & -1 & 0 \\ 0 & 0 & -1 \end{pmatrix}$

D $\begin{pmatrix} -1 & 0 & 0 \\ 0 & 1 & 0 \\ 0 & 0 & 1 \end{pmatrix}$ **E** $\begin{pmatrix} -1 & 0 & 0 \\ 0 & -1 & 0 \\ 0 & 0 & 1 \end{pmatrix}$

The three columns of the matrix represent the images of the three points $I(1,0,0)$, $J(0,1,0)$ and $K(0,0,1)$. In a half-turn about the x-axis, I is invariant (which rules out answers B, D and E), J goes to $(0,-1,0)$ and K goes to $(0,0,-1)$, which rules out answer A.

ANSWER C

Example 5
If matrix X has order $p \times q$ and matrix Y has order $q \times r$ then matrix XY has order

A $p \times q$ **B** $q \times r$ **C** $r \times p$ **D** $p \times r$ **E** $q \times p$

Matrix	X	$\times$	Y	$=$	XY
Order	$p \times q$		$q \times r$		$p \times r$

In order for X and Y to be multiplied together (to be compatible) the number of columns in X (q) must equal the number of rows in Y (q). The product XY will have p rows and r columns, i.e., order $p \times r$.
ANSWER D

Example 6
If matrix X has order $p \times q$, matrix Y has order $q \times r$ and matrix Z has order $r \times s$ then matrix XYZ has order

A $p \times q$ **B** $q \times r$ **C** $r \times s$ **D** $p \times r$ **E** $p \times s$

Matrix multiplication is associative so $(XY)Z = X(YZ)$. From Example 5, the order of XY is $p \times r$, so the order of $(XY)Z$ is $p \times s$. Similarly, the order of YZ is $q \times s$, so the order of $X(YZ)$ is $p \times s$.
ANSWER E

Example 7
Which matrix does not represent an enlargement?

A $\begin{pmatrix} 3 & 0 \\ 0 & 3 \end{pmatrix}$ **B** $\begin{pmatrix} \frac{1}{2} & 0 \\ 0 & \frac{1}{2} \end{pmatrix}$ **C** $\begin{pmatrix} -2 & 0 \\ 0 & -2 \end{pmatrix}$ **D** $\begin{pmatrix} 3 & 0 & 0 \\ 0 & 3 & 0 \\ 0 & 0 & 3 \end{pmatrix}$

E $\begin{pmatrix} -2 & 0 & 0 \\ 0 & -2 & 0 \\ 0 & 0 & -2 \end{pmatrix}$

Matrix A takes (x, y) to $(3x, 3y)$ and multiplies lengths by 3 and areas by 9. A mathematical 'enlargement' can also reduce a figure in size, e.g. matrix B will reduce dimensions by one half and the area by one quarter of the original size. Matrix C is a negative enlargement which in two dimensions is equivalent to a rotation of 180° followed by an enlargement of scale factor $+2$. Matrix D is a three-dimensional enlargement where lengths are tripled, areas multiplied by 9 and volumes by 27. Matrix E, in addition to doubling lengths, quadrupling areas and multiplying volumes by 8, also turns a three-dimensional figure inside out and destroys the sense of the transformation. Matrix E is therefore not an enlargement. ANSWER E

Multiple choice questions (Type B) Answer according to the table

A	**B**	**C**	**D**	**E**
1, 2, 3 correct	1, 3 only	2, 3 only	2 only	3 only

Example 8

The determinant of the matrix $\begin{pmatrix} a & b \\ c & d \end{pmatrix}$ is

1 $ab-cd$ **2** $ac-bd$ **3** $ad-bc$

This is simply definition: statement 3 only is correct. ANSWER E

Example 9

The determinant of the matrix $\begin{pmatrix} a & b & c \\ d & e & f \\ g & h & k \end{pmatrix}$ is equal to

1 $a(ek-fh)+b(dk-fg)+c(dh-ge)$
2 $a(ek-fh)+b(fg-dk)+c(dh-ge)$
3 $a(ek-fh)-b(dk-fg)+c(dh-ge)$

Statement 3 is the form usually remembered and statement 2 is the same as this but slightly rearranged. Statements 2 and 3 are correct. ANSWER C

Example 10

$$\underset{A}{\begin{pmatrix} 1 & 2 \\ 2 & 4 \end{pmatrix}}\underset{B}{\begin{pmatrix} a & b \\ c & d \end{pmatrix}} = \underset{C}{\begin{pmatrix} 1 & 0 \\ 0 & 1 \end{pmatrix}}$$

1 C is the identity matrix. **2** B is the inverse of A.
3 The values of a, b, c and d cannot be found.

Statement 1 is correct: $\begin{pmatrix}1 & 0\\0 & 1\end{pmatrix}\begin{pmatrix}a & b\\c & d\end{pmatrix}=\begin{pmatrix}a & b\\c & d\end{pmatrix}=\begin{pmatrix}a & b\\c & d\end{pmatrix}\begin{pmatrix}1 & 0\\0 & 1\end{pmatrix}$

If the values of a, b, c and d could be found then B would be the inverse of A, but these values cannot be found so A has no inverse.
$a+2c = 1$ and $2a+4c = 0$, which are inconsistent.
Statements 1 and 3 are correct. ANSWER B

In general, a matrix cannot have an inverse if its determinant is zero.

> **Example 11**
> If $AB = I$ where A and B are matrices and I is the 2×2 identity matrix
>
> **1** A and B have the same order. **2** B is the inverse of A.
> **3** A and B are compatible.

Statement 3 means that A and B can be multiplied together (which is true) and this can happen if A is $\begin{pmatrix}1 & 2 & 0\\2 & 5 & 0\end{pmatrix}$ and B is $\begin{pmatrix}5 & -2\\-2 & 1\\3 & 4\end{pmatrix}$, i.e., A and B are of different order.

Only statement 3 is correct. ANSWER E

Short questions

> **Example 12**
> Show that for a transformation in the xy plane represented by the matrix $\begin{pmatrix}a & b\\c & d\end{pmatrix}$ the area is multiplied by $\Delta = ad-bc$, the determinant of the matrix.

In Figure 18(a) the unit square $OIUJ$ (area 1) is transformed into the parallelogram $OI'U'J'$. The coordinates can be evaluated by matrix multiplication and $\mathbf{OJ}' = b\mathbf{i}+d\mathbf{j} = \mathbf{I'U'}$.

$$\begin{aligned}\text{Area of } OI'U'J' &= 2\times \text{area triangle } OI'U'\\ &= 2\times\{\tfrac{1}{2}(a+b)(c+d)-\tfrac{1}{2}ac-bc-\tfrac{1}{2}bd\}\\ &= ac+ad+bc+bd-ac-2bc-bd = ad-bc = \Delta\end{aligned}$$

The area scale factor is $\Delta = ad-bc$.

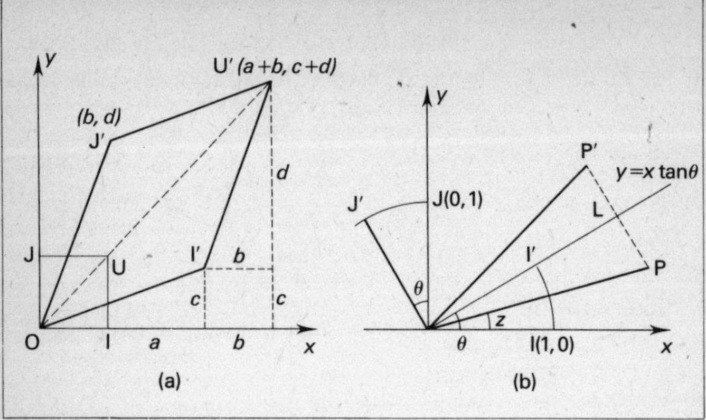

Figure 18

> **Example 13**
> Find the matrix R which represents a rotation of $+\theta°$ about the origin, and the matrix M, which represents a reflexion in the line $y = x \tan\theta$ (both in two dimensions).

With the notation in Figure 18(b) the point $I(1,0)$ rotates through θ to the point $I'(\cos\theta,\sin\theta)$ and $J(0,1)$ rotates to $J'(-\sin\theta,\cos\theta)$

$$R = \begin{pmatrix} \cos\theta & -\sin\theta \\ \sin\theta & \cos\theta \end{pmatrix} \qquad \begin{array}{l} \det R = \cos^2\theta + \sin^2\theta = 1 \\ \text{area is preserved (invariant)} \end{array}$$

For M, consider the image P' of the point $P(r\cos z, r\sin z)$.

The length of $OP = r \Rightarrow OP' = r$ (reflexion)

$\angle POL = \theta - z = \angle LOP'$ so $\angle IOP' = \theta + \theta - z = 2\theta - z$

P' is $[r\cos(2\theta - z), r\sin(2\theta - z)]$

M takes $P(r\cos z, r\sin z)$ to $P'[r\cos(2\theta - z), r\sin(2\theta - z)]$

$$M\begin{pmatrix} r\cos z \\ r\sin z \end{pmatrix} = \begin{pmatrix} r\cos(2\theta - z) \\ r\sin(2\theta - z) \end{pmatrix} = \begin{pmatrix} r\cos 2\theta\cos z + r\sin 2\theta\sin z \\ r\sin 2\theta\sin z - r\cos 2\theta\sin z \end{pmatrix}$$

$$= \begin{pmatrix} \cos 2\theta & \sin 2\theta \\ \sin 2\theta & -\cos 2\theta \end{pmatrix}\begin{pmatrix} r\cos z \\ r\sin z \end{pmatrix}$$

$$M = \begin{pmatrix} \cos 2\theta & \sin 2\theta \\ \sin 2\theta & -\cos 2\theta \end{pmatrix} \qquad \det M = -\cos^2 2\theta - \sin^2 2\theta = -1.$$

The area is invariant, the negative sign indicating a reflexion (rather than a rotation).

Example 14
Demonstrate the result that $\det(AB) = \det A \times \det B$ by finding the determinants of the 2×2 matrices $A = \begin{pmatrix} a & b \\ c & d \end{pmatrix}$, $B = \begin{pmatrix} e & f \\ g & h \end{pmatrix}$ and AB.

$$AB = \begin{pmatrix} a & b \\ c & d \end{pmatrix}\begin{pmatrix} e & f \\ g & h \end{pmatrix} = \begin{pmatrix} ae+bg & af+bh \\ ce+dg & cf+dh \end{pmatrix}$$

$$\begin{aligned}\det(AB) &= (ae+bg)(cf+dh)-(af+bh)(ce+dg) \\ &= aecf+aedh+bgcf+bgdh-afce-afdg-bhce-bhdg \\ &= aedh-adfg-bceh+bgcf = ad(eh-fg)-bc(eh-fg) \\ &= (ad-bc)(eh-fg) \\ &= \det A \times \det B\end{aligned}$$

Example 15

Given $A = \begin{pmatrix} 1 & 2 & 3 \\ 2 & 0 & 1 \end{pmatrix}$ $B = \begin{pmatrix} 2 & 3 \\ -1 & 2 \\ 2 & 1 \end{pmatrix}$ $C = \begin{pmatrix} 1 \\ 3 \end{pmatrix}$ find $(AB)C$ and $A(BC)$ and comment.

$$(AB)C = \left[\begin{pmatrix} 1 & 2 & 3 \\ 2 & 0 & 1 \end{pmatrix}\begin{pmatrix} 2 & 3 \\ -1 & 2 \\ 2 & 1 \end{pmatrix}\right]\begin{pmatrix} 1 \\ 3 \end{pmatrix} = \begin{pmatrix} 6 & 10 \\ 6 & 7 \end{pmatrix}\begin{pmatrix} 1 \\ 3 \end{pmatrix} = \begin{pmatrix} 36 \\ 27 \end{pmatrix}$$

$$A(BC) = \begin{pmatrix} 1 & 2 & 3 \\ 2 & 0 & 1 \end{pmatrix}\left[\begin{pmatrix} 2 & 3 \\ -1 & 2 \\ 2 & 1 \end{pmatrix}\begin{pmatrix} 1 \\ 3 \end{pmatrix}\right] = \begin{pmatrix} 1 & 2 & 3 \\ 2 & 0 & 1 \end{pmatrix}\begin{pmatrix} 11 \\ 5 \\ 5 \end{pmatrix} = \begin{pmatrix} 36 \\ 27 \end{pmatrix}$$

$A(BC) = (AB)C$

This is one example which shows that matrix multiplication is associative.

Example 16
Use the matrix for a rotation of θ to transform the point $(\cos z, \sin z)$ to the point $(\cos(z+\theta), \sin(z+\theta))$ and deduce the addition formulae.

The point $P(\cos z, \sin z)$ is distant 1 from the origin and when rotated through an angle θ to Q, OQ makes an angle $(z+\theta)$ with the x-axis. Since OQ has length 1, the coordinates of Q are

$$(\cos(z+\theta), \sin(z+\theta))$$

and

$$\begin{pmatrix}\cos(z+\theta)\\ \sin(z+\theta)\end{pmatrix} = \begin{pmatrix}\cos\theta & -\sin\theta\\ \sin\theta & \cos\theta\end{pmatrix}\begin{pmatrix}\cos z\\ \sin z\end{pmatrix}$$

$$= \begin{pmatrix}\cos z\cos\theta - \sin z\sin\theta\\ \sin z\cos\theta + \cos z\sin\theta\end{pmatrix}$$

Example 17
Prove the result for matrices $(AB)^{-1} = B^{-1}A^{-1}$.

Knowing that matrix multiplication is associative

$$\begin{aligned} B^{-1}A^{-1}AB &= B^{-1}IB \text{ since } A^{-1}A = I\\ &= B^{-1}B = I \end{aligned}$$

$(AB)^{-1}$ is the inverse of AB so

$$(AB)^{-1}AB = I \Rightarrow (AB)^{-1} = B^{-1}A^{-1}.$$

Alternatively, $(AB)^{-1}AB = I \Rightarrow (AB)^{-1}ABB^{-1} = IB^{-1}$

$$\begin{aligned} &\Rightarrow (AB)^{-1}A = B^{-1}\\ &\Rightarrow (AB)^{-1}AA^{-1} = B^{-1}A^{-1}\\ &\Rightarrow (AB)^{-1} = B^{-1}A^{-1} \end{aligned}$$

Example 18
Find the inverse of $\begin{pmatrix}1 & 2 & 3\\ 2 & 5 & -1\\ 3 & 8 & -4\end{pmatrix}$ and hence or otherwise

solve

$$\begin{aligned} x+2y+3z &= 14\\ 2x+5y-z &= 9\\ 3x+8y-4z &= 7 \end{aligned}$$

Method 1 The inverse of

$$M = \begin{pmatrix}a & b & c\\ d & e & f\\ g & h & i\end{pmatrix} \text{ is } \frac{1}{\Delta}\begin{pmatrix}A & D & G\\ B & E & H\\ C & E & I\end{pmatrix} \text{ where } \begin{pmatrix}A & B & C\\ D & E & F\\ G & H & I\end{pmatrix}$$

is the matrix of the cofactors of M. Each element of M has a cofactor (denoted by the corresponding capital letter) which is the value of the 2×2 determinant left by omitting the row and column containing the letter (remembering to change the sign of B, D, F and H). $\Delta = \det M$, e.g.

$$A = \begin{vmatrix} e & f \\ h & i \end{vmatrix} = ei - fh \qquad B = -\begin{vmatrix} d & f \\ g & i \end{vmatrix} = -(di - fg)$$

$A = -20+8 = -12$; $B = -(-8+3) = 5$; $C = 16-15 = 1$;
$D = -(-8-24) = 32$; $E = -4-9 = -13$; $F = -(8-6) = -2$;
$G = -2-15 = -17$; $H = -(-1-6) = 7$; $I = 5-4 = 1$

$\Delta = \det M = 1(-20+8)-2(-8+3)+3(16-15) = -12+10+3 = 1$

$$M^{-1} = \begin{pmatrix} -12 & 32 & -17 \\ 5 & -13 & 7 \\ 1 & -2 & 1 \end{pmatrix} \text{ and } \begin{pmatrix} x \\ y \\ z \end{pmatrix} = M^{-1}\begin{pmatrix} 14 \\ 9 \\ 7 \end{pmatrix}$$

$$= \begin{pmatrix} -12 & 32 & -17 \\ 5 & -13 & 7 \\ 1 & -2 & 1 \end{pmatrix}\begin{pmatrix} 14 \\ 9 \\ 7 \end{pmatrix} = \begin{pmatrix} 1 \\ 2 \\ 3 \end{pmatrix}$$

Method 2 Solve the equations

$$x+2y+3z = 14 \quad (1)$$
$$2x+5y-z = 9 \quad (2)$$
$$3x+8y-4z = 7 \quad (3)$$

(2) − 2(1) gives $y-7z = -19$ (4)
(3) − 3(1) gives $2y-13z = -35$ (5)
(5) − 2(4) gives $z = 3$

Substituting back in (4) gives $y = 7z-19 = 21-19 = 2 \Rightarrow x = 1$.

Many students go round in circles with this method because they forget to stick to a pattern:

1. Eliminate x from (1) and (2).
2. Eliminate x from (1) and (3).
3. Eliminate y from (4) and (5) to give z, then find y and x.

Method 3 Do the same row operations (as in method 2) to reduce the coefficient matrix to echelon form and then to the identity. If the same operations are done to the identity the inverse is obtained.

$$\begin{pmatrix} 1 & 2 & 3 \\ 2 & 5 & -1 \\ 3 & 8 & -4 \end{pmatrix} \begin{pmatrix} 14 \\ 9 \\ 7 \end{pmatrix} \qquad \begin{pmatrix} 1 & 0 & 0 \\ 0 & 1 & 0 \\ 0 & 0 & 1 \end{pmatrix}$$

Row 2 − 2 × Row 1
Row 3 − 3 × Row 1

$$\begin{pmatrix} 1 & 2 & 3 \\ 0 & 1 & -7 \\ 0 & 2 & -13 \end{pmatrix} \begin{pmatrix} 14 \\ -19 \\ -35 \end{pmatrix} \qquad \begin{pmatrix} 1 & 0 & 0 \\ -2 & 1 & 0 \\ -3 & 0 & 1 \end{pmatrix}$$

$$\text{Row }3-2\times\text{Row }2 \quad \begin{pmatrix}1 & 2 & 3\\0 & -1 & -7\\0 & 0 & 1\end{pmatrix}\begin{pmatrix}14\\-19\\3\end{pmatrix} \quad \begin{pmatrix}1 & 0 & 0\\-2 & 1 & 0\\1 & -2 & 1\end{pmatrix}$$

$$\text{Row }2+7\times\text{Row }3 \quad \begin{pmatrix}1 & 2 & 3\\0 & 1 & 0\\0 & 0 & 1\end{pmatrix}\begin{pmatrix}14\\2\\3\end{pmatrix} \quad \begin{pmatrix}1 & 0 & 0\\5 & -13 & 7\\1 & -2 & 1\end{pmatrix}$$

$$\text{R1}-2\text{R2}-3\text{R3} \quad \begin{pmatrix}1 & 0 & 0\\0 & 1 & 0\\0 & 0 & 1\end{pmatrix}\begin{pmatrix}1\\2\\3\end{pmatrix}\begin{pmatrix}-12 & 32 & -17\\5 & -13 & 7\\1 & -2 & 1\end{pmatrix}$$

Identity Solution Inverse

Example 19
Find the solution of the three equations $x+2y+3z = 4$, $2x+3y+4z = 2$ and $3x+5y+pz = r$ in the cases where

(a) $p = 8$ and $r = 8$ (b) $p = 7$ and $r = 8$ (c) $p = 7$ and $r = 6$

$$\text{(a)} \begin{pmatrix}1 & 2 & 3\\2 & 3 & 4\\3 & 5 & 8\end{pmatrix}\begin{pmatrix}x\\y\\z\end{pmatrix} = \begin{pmatrix}4\\2\\8\end{pmatrix} \Rightarrow \begin{pmatrix}1 & 2 & 3\\0 & -1 & -2\\0 & -1 & -1\end{pmatrix}\begin{pmatrix}x\\y\\z\end{pmatrix} = \begin{pmatrix}4\\-6\\-4\end{pmatrix}$$

$$\Rightarrow \begin{pmatrix}1 & 2 & 3\\0 & -1 & -2\\0 & 0 & 1\end{pmatrix}\begin{pmatrix}x\\y\\z\end{pmatrix} = \begin{pmatrix}4\\-6\\2\end{pmatrix}$$

$\Rightarrow$ third equation is $1\times z = 2 \Rightarrow z = 2$
second equation is $y+2z = 6 \Rightarrow y = 2$
first equation is $x+2y+3z = 4 \Rightarrow x = -6$

The three equations represent three planes which intersect at the point $(-6,2.2)$.

$$\text{(b)} \begin{pmatrix}1 & 2 & 3\\2 & 3 & 4\\3 & 5 & 7\end{pmatrix}\begin{pmatrix}x\\y\\z\end{pmatrix} = \begin{pmatrix}4\\2\\8\end{pmatrix} \quad \begin{pmatrix}1 & 2 & 3\\0 & -1 & -2\\0 & -1 & -2\end{pmatrix}\begin{pmatrix}x\\y\\z\end{pmatrix} = \begin{pmatrix}4\\-6\\-4\end{pmatrix}$$

$$\begin{pmatrix}1 & 2 & 3\\0 & -1 & -2\\0 & 0 & 0\end{pmatrix}\begin{pmatrix}x\\y\\z\end{pmatrix} = \begin{pmatrix}4\\-6\\2\end{pmatrix}$$

$\Rightarrow$ third equation is $0 \times z = 2 \Rightarrow$ **No solution**.
The planes do not intersect: the planes are not parallel, but the line of intersection of any two is parallel to the third plane. The planes are said to form a prism.

(c) $\begin{pmatrix} 1 & 2 & 3 \\ 2 & 3 & 4 \\ 3 & 5 & 7 \end{pmatrix} \begin{pmatrix} x \\ y \\ z \end{pmatrix} = \begin{pmatrix} 4 \\ 2 \\ 6 \end{pmatrix}$ $\begin{pmatrix} 1 & 2 & 3 \\ 0 & -1 & -2 \\ 0 & -1 & -2 \end{pmatrix} \begin{pmatrix} x \\ y \\ z \end{pmatrix} = \begin{pmatrix} 4 \\ -6 \\ -6 \end{pmatrix}$

$$\begin{pmatrix} 1 & 2 & 3 \\ 0 & -1 & -2 \\ 0 & 0 & 0 \end{pmatrix} \begin{pmatrix} x \\ y \\ z \end{pmatrix} = \begin{pmatrix} 4 \\ -6 \\ 0 \end{pmatrix}$$

$\Rightarrow$ third equation is $0 \times z = 0 \Rightarrow z =$ any value.

Put $z = t \Rightarrow y = 6 - 2t$ from the second equation $y + 2z = 6$.
First equation is $x + 2y + 3z = 4 \Rightarrow x = 4 - 12 + 4t - 3t = t - 8$

The solution is $\begin{pmatrix} x \\ y \\ z \end{pmatrix} = t \begin{pmatrix} 1 \\ -2 \\ 1 \end{pmatrix} + \begin{pmatrix} -8 \\ 6 \\ 0 \end{pmatrix}$

which represents a straight line through the point $(-8,6.0)$ in the direction $\mathbf{i} + 2\mathbf{j} + \mathbf{k}$, i.e., the line joining the origin to $(1,-2.1)$. The three planes meet in a line, not a single point.

Chapter 8
Series

This section of the syllabuses covers the various types of series and sequences and the methods for summing them. Most syllabuses mention the arithmetical and geometrical progressions and the binomial theorem but some include the expansions of trigonometrical, exponential and logarithmic functions.

The method of induction and a limited knowledge of convergence are required as well as the techniques needed for selection (called permutations and combinations).

Multiple choice questions (Type A) Select the correct answer

> **Example 1**
> The sum of the first 10 terms of the series $16+13+10+7+\ldots$ is
>
> **A** 10 **B** 20 **C** 25 **D** 50 **E** 77

The series is an arithmetical progression since the difference of successive terms is constant. The common difference, d, equals -3.

The sum to n terms $S_n = \frac{n}{2}[2a+(n-1)d]$ where a is the first term, n is the number of terms and d is the common difference.

$\therefore\ S_{10} = \frac{10}{2}[32+9(-3)] = 25$ ANSWER C

> **Example 2**
> The sum to infinity of the series $0.03+0.003+0.0003+\ldots$ is
>
> **A** $\frac{1}{3}$ **B** $\frac{1}{30}$ **C** $\frac{1}{33}$ **D** $\frac{10}{9}$ **E** infinity

This series is a geometrical progression with a common ratio of $r = \frac{1}{10}$ and a first term of $\frac{3}{100}$.

The sum to n terms $S_n = \dfrac{a(1-r^n)}{1-r}$

However, since $-1 < r < 1$, $r^n \to 0$ as $n \to \infty$.

$\therefore\ S_\infty = \dfrac{a}{1-r} = \dfrac{3}{100}\left[\dfrac{1}{1-\frac{1}{10}}\right] = \frac{3}{100} \times \frac{10}{9} = \frac{1}{30}$ ANSWER B

Example 3
The range of validity of the expansion

$$\ln\left(\frac{1+x}{1-x}\right) = 2x+\frac{2x^3}{3}+\ldots+\frac{2x^{2n-1}}{(2n-1)}+\ldots \text{ is}$$

A $-1 < x \leqslant 1$ **B** $-1 \leqslant x < 1$ **C** $-1 \leqslant x \leqslant 1$
D $-1 < x < 1$ **E** all x

The given function is a combination of two logarithmic functions. Thus, since $\ln\left(\frac{1+x}{1-x}\right) = \ln(1+x) - \ln(1-x)$ and $\ln(1+x)$ is valid for $-1 < x \leqslant 1$ and $\ln(1-x)$ is valid for $-1 \leqslant x < 1$, the given expansion will only be valid when both of these conditions hold, i.e., when $-1 < x < 1$.
ANSWER D

Example 4
The first 3 terms of the expansion of $(9-2x)^{1/2}$ in ascending powers of x are

A $3+\frac{1}{3}x+\frac{1}{54}x^2$ **B** $3-\frac{1}{3}x+\frac{1}{54}x^2$ **C** $3-\frac{1}{3}x-\frac{1}{54}x^2$
D $3-\frac{1}{3}x+\frac{1}{12}x^2$ **E** $3-\frac{1}{3}x^2-\frac{1}{12}x^2$

Now $(9-2x)^{1/2} = 9^{1/2}(1-\frac{2}{9}x)^{1/2} = 3(1-\frac{2}{9}x)^{1/2}$
Using the binomial expansion for a rational index

$$3(1-\tfrac{2}{9}x)^{1/2} = 3\left[1-\tfrac{1}{2}(\tfrac{2}{9}x) + \frac{(\frac{1}{2})(-\frac{1}{2})}{2!}(-\tfrac{2}{9}x)^2+\ldots\right]$$

$$= 3[1-\tfrac{1}{9}x-\tfrac{1}{162}x^2+\ldots] = 3-\tfrac{1}{3}x-\tfrac{1}{54}x^2+\ldots$$

ANSWER C

Example 5
The number of different arrangements that can be made by using all the letters of the word MINIMUM is

A $7!$ **B** $\frac{7!}{2}$ **C** $\frac{7!}{3}$ **D** $\frac{7!}{6}$ **E** $\frac{7!}{12}$

If the letters were all different the number of arrangements would be 7!. However, in this case the letter M occurs three times and these can be arranged amongst themselves in 3! ways without altering the positions of the other letters.

Similarly, the letter I occurs twice and these can be arranged in 2! ways.

Hence the total number of distinct arrangements $= \frac{7!}{3!2!} = \frac{7!}{12}$.

ANSWER E

Example 6
The number of ways in which 3 books can be chosen from a shelf containing 8 different books is

A $8!$ **B** $3!$ **C** $\frac{8!}{3!}$ **D** $\frac{8!}{5!}$ **E** $\frac{8!}{5!3!}$

Since the order in which the 3 books are chosen is not important the question is asking for the number of different groups of 3 books that can be chosen, i.e., a combination. The number of ways 3 items can be chosen from 8 (where the order is important) is $\frac{8!}{5!}$. However, each group of 3 books can be rearranged amongst themselves in 3! ways.

$\therefore$ the number of distinct groups of 3 books $= \frac{8!}{5!3!}$. ANSWER E

Note that if you can remember that the number of ways of choosing r items from n different items, where the order is not important, is ${}^nC_r = \frac{n!}{(n-r)!r!}$ then the solution can be found quickly.

Multiple choice questions (Type B) Answer according to the table

A	B	C	D	E
1, 2, 3 correct	1, 3 only	2, 3 only	2 only	3 only

Example 7
If $(1+x)^n = a_0 + a_1x + a_2x^2 + \ldots + a_rx^r + \ldots + a_nx^n$ where n is a positive integer

1 $a_r = \frac{n(n-1)\ldots(n-r-1)}{r!}$

2 $a_0 + a_1 + a_2 + \ldots + a_n = 2^n$

3 Sum of the even coefficients = sum of the odd coefficients.

The binomial expansion for a positive integral index gives

$$(1+x)^n = 1+{}^nC_1x+{}^nC_2x^2+\ldots+{}^nC_rx^r+\ldots+x^n$$

$$\therefore \text{ the coefficient of } x^r \text{ is } a_r = {}^nC_r = \frac{n!}{(n-r)!\,r!}$$

$$= \frac{n(n-1)(n-2)\ldots(n-r+1)}{r!}$$

Statement 1 is incorrect.
Substituting $x = 1$ in the expansion we obtain

$$a_0+a_1+a_2+\ldots+a_n = 2^n$$

Statement 2 is correct.

Substituting $x = -1$, $\quad 0 = a_0-a_1+a_2-a_3+\ldots+(-1)^na_n$

Hence, if n is even

$$a_1+a_3+a_5+\ldots+a_{n-1} = a_0+a_2+a_4+\ldots+a_n$$

If n is odd

$$a_1+a_3+a_5+\ldots+a_n = a_0+a_2+a_4+\ldots+a_{n-1}$$

In either case statement 3 is correct.
Statements 2 and 3 only are true. ANSWER C

Example 8
If $e^{x^2} = 1+a_1x+a_2x^2+\ldots$

1 $a_r = 0$ for odd values of r **2** $a_2 = 1$

3 The general term is $\dfrac{x^{2r}}{r!}$

The expansion of $e^x = 1+x+\dfrac{x^2}{2!}+\dfrac{x^3}{3!}+\ldots$

Replacing x by x^2 we obtain

$$e^{x^2} = 1+x^2+\frac{x^4}{2!}+\frac{x^6}{3!}+\ldots+\frac{x^{2r}}{r!}+\ldots$$

By comparing with the given expansion it can be seen that $a_r = 0$ for all odd values of r, $a_2 = 1$ and the general term is as given.
Statements 1, 2 and 3 are all true. ANSWER A

General questions

In this section, questions tend to require the expansions of one of the functions on the syllabus by the binomial theorem or by another standard method.

Example 9
If $(1+\frac{3}{2}x)^{-1/2} = 1-\frac{3}{4}x+ax^2+bx^3$ as far as the term in x^3, find a and b.

Using the binomial theorem

$$(1+\tfrac{3}{2}x)^{-1/2} = 1+(-\tfrac{1}{2})(\tfrac{3}{2}x)+\frac{(-\frac{1}{2})(-\frac{3}{2})}{2!}(\tfrac{3}{2}x)^2 + \frac{(-\frac{1}{2})(-\frac{3}{2})(-\frac{5}{2})}{3!}(\tfrac{3}{2}x)^3+\dots$$

$$= 1-\tfrac{3}{4}x+\tfrac{27}{32}x^2-\tfrac{135}{128}x^3+\dots$$

Hence $a = \frac{27}{32}$ and $b = -\frac{135}{128}$.

Example 10
Use the expansion of $(1-x)^{1/2}$ to estimate the value of $\sqrt{11}$ to 4 decimal places.

Expanding by the binomial theorem

$$(1-x)^{1/2} = 1-\tfrac{1}{2}x+\frac{(\frac{1}{2})(-\frac{1}{2})}{2!}(-x)^2+\frac{(\frac{1}{2})(-\frac{1}{2})(-\frac{3}{2})}{3!}(-x)^3+\dots$$

$$= 1-\tfrac{1}{2}x-\tfrac{1}{8}x^2-\tfrac{1}{16}x^3-\dots$$

This is valid if $-1 < x < 1$. Let $x = 0.01$.

$\therefore\ (1-0.01)^{1/2} = 1-\frac{1}{2}(0.01)-\frac{1}{8}(0.01)^2-\frac{1}{16}(0.01)^3-\dots$

$\therefore\ \dfrac{\sqrt{99}}{10} \simeq 1-0.005-0.000\,012\,5 = 0.994\,987\,5$

Since $\dfrac{\sqrt{99}}{10} = \dfrac{3\sqrt{11}}{10} \Rightarrow \sqrt{11} \simeq \frac{10}{3}\times 0.994\,987\,5 = 3.316\,625.$

$\therefore\ \sqrt{11} = 3.3166$ correct to 4 decimal places.

Example 11
Find the coefficient of x^n in the expansion of $(a+bx)(1-2x)^{-1}$.

Expanding $(1-2x)^{-1}$ by the binomial theorem

$$(1-2x)^{-1} = 1+2x+\frac{(-1)(-2)}{2!}(-2x)^2$$

$$+\frac{(-1)(-2)(-3)}{3!}(-2x)^3+\ldots$$

$\therefore\ (a+bx)(1-2x)^{-1} = (a+bx)[1+2x+(2x)^2+\ldots+(2x)^r+\ldots]$

The term in x^n can be obtained by multiplying out the bracket choosing only the terms in x^{n-1} and x^n from the square bracket.

The term in $x^n = a(2x)^n+bx(2x)^{n-1} = 2^{n-1}(2a+b)x^n$.

The coefficient of the term in x^n is $(2a+b)2^{n-1}$.

> **Example 12**
>
> Find the coefficient of x in the expansion of $\left(\frac{x^2}{2}-\frac{2}{x}\right)^8$.

The general term of the expansion $(a+b)^n$ is ${}^nC_r a^{n-r}b^r$.

In this case the general term is ${}^8C_r\left(\frac{x^2}{2}\right)^{8-r}\left(-\frac{2}{x}\right)^r$.

If this is to produce a term in x then the index must be 1, i.e.

$$2(8-r)-r = 1 \Rightarrow 3r = 15 \Rightarrow r = 5$$

The term in x is ${}^8C_5\left(\frac{x^2}{2}\right)^3\left(-\frac{2}{x}\right)^5 = {}^8C_5 \times \frac{(-2)^5}{2^3}x$.

The coefficient of x is $-4\times{}^8C_5 = -4 \times \frac{8!}{5!3!} = -224$.

> **Example 13**
>
> (i) Find the number of ways a team of 4 can be chosen from 5 boys and 3 girls if
> (a) it must contain 2 boys and 2 girls,
> (b) it must contain at least 1 boy and 1 girl.
>
> (ii) Find the sum to infinity of the series $\frac{2}{1!}+\frac{3}{2!}+\frac{4}{3!}+\ldots$

(i) (a) The boys can be chosen in ${}^5C_2$ ways $= \frac{5}{3!2!}$ ways $= 10$ ways.

The girls can be chosen in ${}^3C_2$ ways $= \frac{3}{2!1!}$ ways $= 3$ ways.

The total number of teams $= 10\times 3 = 30$.

(i) (b) If the team must contain at least 1 boy and 1 girl it can be formed in the following ways.

1 boy and 3 girls $= {}^5C_1 \times {}^3C_3 = 5 \times 1 = 5$ ways

2 boys and 2 girls $= 30$ ways (from part (a))

3 boys and 1 girl $= {}^5C_3 \times {}^3C_1 = 10 \times 3 = 30$ ways

The total number of teams $= 5+30+30 = 65$.

(ii) The series $\frac{2}{1!}+\frac{3}{2!}+\frac{4}{3!}+\ldots$ can be written $\sum_{r=1}^{\infty} \frac{r+1}{r!}$

But $$\sum_{r=1}^{\infty} \frac{r+1}{r!} = \sum_{r=1}^{\infty}\left(\frac{1}{(r-1)!}+\frac{1}{r!}\right)$$

Consider $$e^x = 1+x+\frac{x^2}{2!}+\frac{x^3}{3!}+\ldots+\frac{x^r}{r!}+\ldots$$

when $x = 1$ $$e = 1+1+\frac{1}{2!}+\frac{1}{3!}+\ldots+\frac{1}{r!}+\ldots$$

i.e. $\sum_{r=1}^{\infty}\frac{1}{r!} = e-1$ and $\sum_{r=1}^{\infty}\frac{1}{(r-1)!} = e$

$$\therefore \sum_{r=1}^{\infty}\frac{r+1}{r!} = e+(e-1) = 2e-1$$

Example 14

Find the sum to n terms of the series

$$\frac{2}{1\times3\times5}+\frac{3}{3\times5\times7}+\frac{4}{5\times7\times9}+\ldots$$

Deduce the sum to infinity.

The series is given by $\sum_{r=1}^{n}\frac{r+1}{(2r-1)(2r+1)(2r+3)}$

Consider

$$\frac{r+1}{(2r-1)(2r+1)(2r+3)} \equiv \frac{A}{(2r-1)}+\frac{B}{(2r+1)}+\frac{C}{(2r+3)}$$

$$\equiv \frac{A(2r+1)(2r+3)+B(2r-1)(2r+3)+C(2r-1)(2r+1)}{(2r-1)(2r+1)(2r+3)}$$

Equating coefficients

$$r+1 \equiv A(2r+1)(2r+3)+B(2r-1)(2r+3)+C(2r-1)(2r+1)$$

Let $r = \frac{1}{2} \Rightarrow \frac{3}{2} = 8A \Rightarrow A = \frac{3}{16}$

$r = -\frac{1}{2} \Rightarrow \frac{1}{2} = -4B \Rightarrow B = -\frac{1}{8}$

$r = -\frac{3}{2} \Rightarrow -\frac{1}{2} = 8C \Rightarrow C = -\frac{1}{16}$

$$\therefore \sum_{r=1}^{n} \frac{r+1}{(2r-1)(2r+1)(2r+3)} = \frac{1}{16}\left(\frac{3}{2r-1} - \frac{2}{2r+1} - \frac{1}{2r+3}\right)$$

$$\begin{aligned} = \frac{1}{16}\Bigg(& 3 - \tfrac{2}{3} - \tfrac{1}{5} \\ & \tfrac{3}{3} - \tfrac{2}{5} - \tfrac{1}{7} \\ & \tfrac{3}{5} - \tfrac{2}{7} - \tfrac{1}{9} \\ & \tfrac{3}{7} - \tfrac{2}{9} - \tfrac{1}{11} \\ & \cdots\cdots\cdots\cdots \\ & \cdots\cdots\cdots\cdots \\ & \frac{3}{2n-5} - \frac{2}{2n-3} - \frac{1}{2n-1} \\ & \frac{3}{2n-3} - \frac{2}{2n-1} - \frac{1}{2n+1} \\ & \frac{3}{2n-1} - \frac{2}{2n+1} - \frac{1}{2n+3}\Bigg) \end{aligned}$$

It can be seen that there are groups of three fractions which have a sum of zero as shown.

$$\text{Hence } S_n = \tfrac{1}{16}\left(3 - \tfrac{2}{3} + \tfrac{3}{3} - \frac{1}{2n+1} - \frac{2}{2n+1} - \frac{1}{2n+3}\right)$$

$$= \tfrac{1}{16}\left(\tfrac{10}{3} - \frac{3}{2n+1} - \frac{1}{2n+3}\right)$$

$$\therefore \sum_{r=1}^{n} \frac{r+1}{(2r-1)(2r+1)(2r+3)} = \frac{5}{24} - \frac{1}{16}\left(\frac{3}{2n+1} + \frac{1}{2n+3}\right)$$

Clearly as $n \to \infty$ the terms $\dfrac{3}{2n+1}$ and $\dfrac{1}{2n+3} \to 0$.

$$\text{Hence } \sum_{r=1}^{\infty} \frac{r+1}{(2r-1)(2r+1)(2r+3)} = \tfrac{5}{24}.$$

Chapter 9
Complex Numbers

The invention of a complex number was due to Gauss and it enabled mathematicians to proceed further in finding roots of an equation of any degree. Knowledge of the algebraic form, $x+iy$, and the polar form, $r(\cos\theta + i\sin\theta)$, are expected by most of the syllabuses and candidates will be required to form the sum, product and quotient of two complex numbers. Representation of complex numbers on an Argand diagram and use of the diagram are also required.

One important result is De Moivre's theorem although the extent to which it is included depends upon the syllabus.

Multiple choice questions (Type A) Select the correct answer

Example 1
The roots of the equation $x^2+2x+4=0$ are

A $\frac{1}{2}(-2\pm 3\sqrt{2}i)$ **B** $\frac{1}{2}(2\pm 3\sqrt{2}i)$ **C** $-1\pm\sqrt{3}i$
D $1\pm\sqrt{3}i$ **E** none

Using the formula solution for the equation $ax^2+bx+c=0$

$$x=\frac{-b\pm\sqrt{b^2-4ac}}{2a} \quad\Rightarrow\quad x=\frac{-2\pm\sqrt{(-2)^2-4(1)(4)}}{2}$$

$$\therefore\ x=\frac{-2\pm\sqrt{-12}}{2}=\frac{-2\pm 2\sqrt{3}i}{2}=-1\pm\sqrt{3}i \qquad \text{ANSWER C}$$

Note that answer E is not correct for, although there are no real roots, complex roots do exist.

Example 2

$$\frac{3+2i}{2-i}=$$

A $\frac{4+7i}{3}$ **B** $\frac{4+7i}{5}$ **C** $\frac{8+i}{3}$ **D** $\frac{8+i}{5}$ **E** $\frac{8-i}{5}$

Multiply numerator and denominator by the complex conjugate of the denominator, i.e., by $2+i$:

$$\frac{(3+2i)(2+i)}{(2-i)(2+i)} = \frac{6+3i+4i+2i^2}{4-i^2} = \frac{4+7i}{5} \qquad \text{ANSWER B}$$

Example 3
$(3-2i)^2 =$

A $5-6i$ **B** $13-12i$ **C** $13-6i$ **D** $5-12i$ **E** 5

$$(3-2i)^2 = (3-2i)(3-2i) = 9-6i-6i+4i^2 = 5-12i$$

ANSWER D

Example 4
The modulus-argument (polar) form of $-2+2\sqrt{3}i$ is

A $4(\cos\frac{2}{3}\pi + i\sin\frac{2}{3}\pi)$ **B** $4(\cos\frac{5}{6}\pi + i\sin\frac{5}{6}\pi)$
C $-4(\cos\frac{1}{3}\pi + i\sin\frac{1}{3}\pi)$ **D** $\frac{1}{4}(\cos\frac{2}{3}\pi + i\sin\frac{2}{3}\pi)$
E $\frac{1}{4}(\cos\frac{5}{6}\pi + i\sin\frac{5}{6}\pi)$

Now $-2+2\sqrt{3}i = 4(-\frac{1}{2}+\frac{1}{2}\sqrt{3}i) = 4(\cos\frac{2}{3}\pi + i\sin\frac{2}{3}\pi)$

ANSWER A

Example 5
In Figure 19 the points P_1 and P_2 represent the complex numbers z_1 and z_2. The point P_3 represents the complex number

A z_1-z_2 **B** z_1+z_2 **C** z_2-z_1 **D** z_1z_2 **E** z_1/z_2

The sum of two complex numbers z_1 and z_2 represented by the points P_1 and P_2 is found geometrically as the fourth vertex, R, of a parallelogram OP_1P_2R.

Since

$$z_1-z_2 = z_1+(-z_2)$$

it follows that since P_4 represents $(-z_2)$ then P_3 represents z_1-z_2 as it is the fourth vertex of the completed parallelogram $OP_1P_4P_3$. ANSWER A

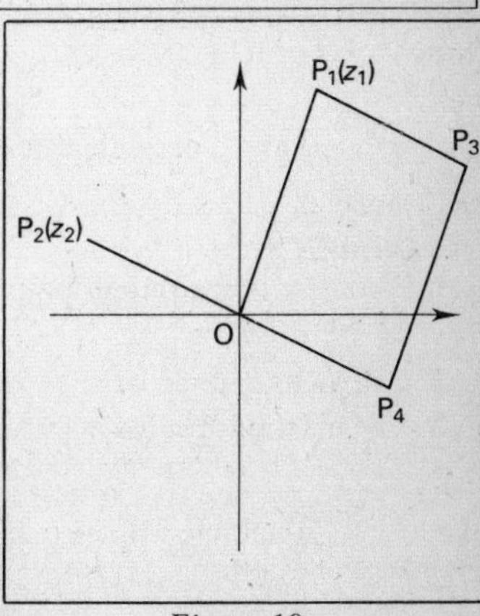

Figure 19

Multiple choice questions (Type B) Answer according to the table

A	B	C	D	E
1, 2, 3 correct	1, 3 only	2, 3 only	2 only	3 only

Example 6

If $z_1 = 1+i$ and $z_2 = 1-\sqrt{3}i$ are two complex numbers

1 $|z_1 z_2| = \dfrac{1}{\sqrt{2}}$ **2** $\arg(z_1 z_2) = -\frac{\pi}{12}$ **3** $|z_1 - z_2| = 1+\sqrt{3}$

Changing z_1 and z_2 into polar form

$$z_1 = 1+i = \sqrt{2}\left(\frac{1}{\sqrt{2}} + \frac{1}{\sqrt{2}}i\right) = \sqrt{2}\,(\cos\tfrac{\pi}{4} + i\sin\tfrac{\pi}{4})$$

$$z_2 = 1-\sqrt{3}i = 2\left(\frac{1}{2} - \frac{\sqrt{3}}{2}i\right) = 2[\cos(-\tfrac{\pi}{3}) + i\sin(-\tfrac{\pi}{3})]$$

Hence $z_1 z_2 = 2\sqrt{2}\,[\cos(\frac{\pi}{4}-\frac{\pi}{3}) + i\sin(\frac{\pi}{4}-\frac{\pi}{3})]$ since the product of z_1 and z_2 is obtained by taking the product of the moduli and the sum of the arguments.

$\therefore\ z_1 z_2 = 2\sqrt{2}\,[\cos(-\frac{\pi}{12}) + i\sin(-\frac{\pi}{12})]$

Statement 1 is incorrect since the modulus is $2\sqrt{2}$.
Statement 2 is correct as the argument is $-\frac{\pi}{12}$.
Now $z_1 - z_2 = (1+\sqrt{3})i \Rightarrow |z_1 - z_2| = 1+\sqrt{3}$ and statement 3 is correct.

Statements 2 and 3 only are true. ANSWER C

Example 7

If $z = x+iy$ is a variable complex number

1 $z\bar{z} = x^2+y^2$.
2 The locus defined by $z\bar{z} = k$ where k is a constant is a circle.
3 The maximum value of $|z|$ such that $|z-(1+i)| \leqslant 1$ is $1+\sqrt{2}$.

If $z = x+iy$ then its complex conjugate $\bar{z} = x-iy$.
Hence $z\bar{z} = (x+iy)(x-iy) = x^2+y^2$.

Statement 1 is correct.

The locus defined by $z\bar{z} = k$ is equivalent to $x^2+y^2 = k$ which represents a circle, centre $(0, 0)$ and radius $\sqrt{k}$.

Statement 2 is correct.

If $|z-(1+i)| \leqslant 1$, then the distance between z and the point representing $1+i$ is $\leqslant 1$, i.e., the point representing z lies on or within the circle, centre $(1, 1)$ and radius 1. The maximum value of $|z|$ is obtained when this point is furthest from the origin.

Therefore, the maximum value of $|z|$ is $1+\sqrt{2}$, since the centre of the circle $(1, 1)$ is a distance $\sqrt{2}$ from the origin. The total distance is therefore $\sqrt{2}+$(the radius of the circle). Notice that the minimum value of $|z|$ will be $\sqrt{2}-1$.

Statement 3 is correct.

Statements 1, 2 and 3 are all true. ANSWER A

General questions

On many examination papers the questions on complex numbers consist of several parts testing different aspects of the topic.

Example 8

If $z_1 = 2+3i$ and $z_2 = 2-i$ express

(a) z_1-2z_2 (b) $z_1^2+z_2^2$ (c) z_1/z_2 in the form $a+bi$.

(a) $z_1-2z_2 = 2+3i-2(2-i) = -2+5i$

(b) $$\begin{aligned} z_1^2+z_2^2 &= (2+3i)^2+(2-i)^2 \\ &= 4+12i+3i^2+4-4i+i^2 = 4+8i \end{aligned}$$

(c) $$\begin{aligned} \frac{z_1}{z_2} = \frac{2+3i}{2-i} = \frac{(2+3i)(2+i)}{(2-i)(2+i)} &= \frac{4+8i+3i^2}{5} \\ &= \frac{1+8i}{5} \end{aligned}$$

Example 9

Find the square roots of $1+\sqrt{3}i$.

Now $1+\sqrt{3}i = 2\left(\frac{1}{2}+\frac{\sqrt{3}}{2}i\right) = 2[\cos(\frac{\pi}{3}+2k\pi)+i\sin(\frac{\pi}{3}+2k\pi)]$

Using De Moivre's theorem

$$(1+\sqrt{3}i)^{1/2} = 2^{1/2}[\cos(\tfrac{\pi}{3}+2k\pi)+i\sin(\tfrac{\pi}{3}+2k\pi)]^{1/2}$$
$$= 2^{1/2}[\cos(\tfrac{\pi}{6}+k\pi)+i\sin(\tfrac{\pi}{6}+k\pi)]$$

If $k = 0$, the root is $\sqrt{2}(\cos\frac{\pi}{6}+i\sin\frac{\pi}{6}) = \dfrac{1}{\sqrt{2}}(\sqrt{3}+i)$.

If $k = 1$, the root is $\sqrt{2}(\cos\frac{7\pi}{6}+i\sin\frac{7\pi}{6}) = -\dfrac{1}{\sqrt{2}}(\sqrt{3}+i)$.

The square roots of $1+\sqrt{3}i$ are $\pm\dfrac{1}{\sqrt{2}}(\sqrt{3}+i)$.

Alternatively, this can be solved by assuming that the root is of the form $a+bi$.

$\therefore\ (a+bi)^2 = 1+\sqrt{3}i \Rightarrow a^2-b^2 = 1$ and $2ab = \sqrt{3}$.

Solving $\quad a^2 - \dfrac{3}{4a^2} = 1 \Rightarrow 4a^4-4a^2-3 = 0$

$$(2a^2+1)(2a^2-3) = 0 \Rightarrow a^2 = \tfrac{3}{2} \text{ or } -\tfrac{1}{2}$$

Since a is real $\quad a = \pm\sqrt{\tfrac{3}{2}} \quad b = \pm\sqrt{\tfrac{1}{2}}$

The square roots are $\pm\dfrac{1}{\sqrt{2}}(\sqrt{3}+i)$.

Example 10
Solve the equation $z^3+8=0$ and show the roots on an Argand diagram.

If $f(z) = z^3+8$ then $f(-2) = 0$ and hence by the factor theorem $z+2$ is a factor.

Now $z^3+8 = (z+2)(z^2-2z+4) = 0$

$\therefore\ z = -2$ or $z^2-2z+4 = 0$

If $z^2-2z+4 = 0$ then $z = \dfrac{2\pm\sqrt{4-16}}{2} = 1\pm i\sqrt{3}$

Hence the roots are $-2, 1\pm i\sqrt{3}$ which are shown in Figure 20(b).

Notice that the roots lie on a circle, centre $(0, 0)$ and radius 2, and are equally spaced around the circle. As usual, the complex roots occur as complex conjugate pairs which must happen if the

coefficients of the original equation are real. In this case, $1+i\sqrt{3}$ and $1-i\sqrt{3}$. In general the use of De Moivre's theorem provides an alternative method which may be necessary for equations of higher degree.

Example 11
If P is the point representing the complex number z on the Argand diagram and $|z| = 2|z+3i|$ find the equation of the locus of P. Sketch the locus of P.

Let $z = x+iy$ and since $|z| = 2|z+3i| \Rightarrow |z|^2 = 4|z+3i|^2$ we have

$$x^2+y^2 = 4[x^2+(y+3)^2]$$

$$x^2+y^2 = 4x^2+4y^2+24y+36$$

$$3x^2+3y^2+24y+36 = 0 \Rightarrow x^2+y^2+8y+12 = 0$$

This can be written as $x^2+(y+4)^2 = 4$ which represents a circle, centre $(0, -4)$ and radius 2. The sketch is shown in Figure 20(a).

This is a particular example of the general result which states that if A and B are two fixed points, the locus of a point P, which moves such that $PA:PB$ is constant, is a circle.
The circle is called the circle of Apollonius.

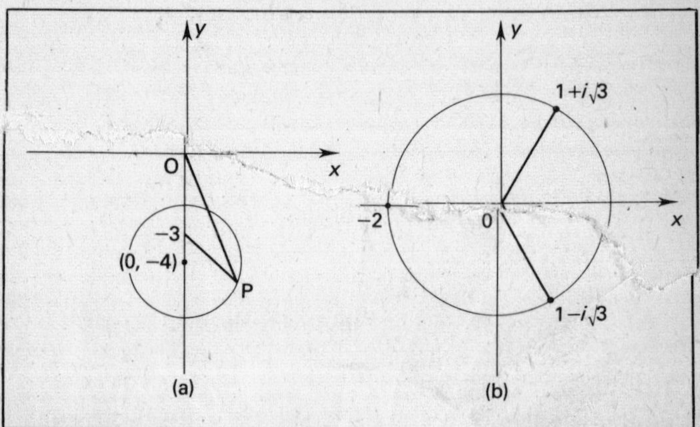

Figure 20

Example 12
Given that one root of the equation $z^3+az^2+bz-15 = 0$ is $2+i$, find the values of a and b and solve the equation if a and b are real.

Since the coefficients of the equation are real any complex roots must occur in complex conjugate pairs. Thus if $2+i$ is a root so is $2-i$.

Hence $[z-(2+i)]$ and $[z-(2-i)]$ are factors.

Since $[z-(2+i)][z-(2-i)] = z^2-4z+5$ we have

$z^3+az^2+bz-15 = (z^2-4z+5)(z-p) = 0$ where the real root is $z = p$.

Equating the coefficients we obtain

constant term	$-5p = -15$	$\Rightarrow p = 3$
coefficient of z^2	$-4-p = a$	$\Rightarrow a = -7$
coefficient of z	$4p+5 = b$	$\Rightarrow b = 17$

The values required are $a = -7$, $b = 17$ and the roots of the equation $z^3-7z^2+17z-15 = 0$ are $z = 3, z = 2 \pm i$.

Example 13
Use De Moivre's theorem to express $\sin^5\theta$ in terms of the sum of the cosines of multiples of θ.

Let $z = \cos\theta + i\sin\theta \Rightarrow \dfrac{1}{z} = z^{-1} = \cos\theta - i\sin\theta$

So $z^n = (\cos\theta + i\sin\theta)^n = \cos n\theta + i\sin n\theta$ and

$$\frac{1}{z^n} = (\cos\theta - i\sin\theta)^n = \cos n\theta - i\sin n\theta$$

$$\text{Hence } (2i\sin\theta)^5 = \left(z - \frac{1}{z}\right)^5$$

$$32i\sin^5\theta = \left(z^5 - \frac{1}{z^5}\right) - 5\left(z^3 - \frac{1}{z^3}\right) + 10\left(z - \frac{1}{z}\right)$$

$$= 2i(\sin 5\theta - 5\sin 3\theta + 10\sin\theta)$$

$$\therefore \quad \sin^5\theta = \tfrac{1}{16}(\sin 5\theta - 5\sin 3\theta + 10\sin\theta)$$

Example 14
Use De Moivre's theorem to express $\sin 5\theta/\sin\theta$ as the sum of powers of $\cos\theta$.

By De Moivre's theorem $\cos 5\theta + i\sin 5\theta = (\cos\theta + i\sin\theta)^5$

Hence $\cos 5\theta + i\sin 5\theta = \cos^5\theta + 5i\cos^4\theta\sin\theta + 10i^2\cos^3\theta\sin^2\theta$
$+ 10i^3\cos^2\theta\sin^3\theta + 5i^4\cos\theta\sin^4\theta + i^5\sin^5\theta$

Equating imaginary parts

$$\sin 5\theta = 5\cos^4\theta\sin\theta - 10\cos^2\theta\sin^3\theta + \sin^5\theta$$

$$\therefore \quad \frac{\sin 5\theta}{\sin\theta} = 5\cos^4\theta - 10\cos^2\theta\sin^2\theta + \sin^4\theta$$

$$= 5\cos^4\theta - 10\cos^2\theta(1-\cos^2\theta) + (1-\cos^2\theta)^2$$

$$= 16\cos^4\theta - 12\cos^2\theta + 1$$

The following example illustrates how the different aspects of the work on complex numbers can be woven together to develop an idea.

Example 15

The complex numbers $z_1 = \dfrac{\sqrt{2}(1-3i)}{2-i}$ and $z_2 = (1+i)^2$ are represented by the points P_1 and P_2 respectively on the Argand diagram.

(i) Express z_1 and z_2 in the form $a+bi$.

(ii) Show that OP_1P_2 is an isosceles triangle if O is the origin.

(iii) Find the locus of the point P representing the complex number z if $|z-z_1| = |z-z_2|$.

(iv) Describe the locus of the point Q which represents the complex number z if $\arg\left(\dfrac{z-z_1}{z-z_2}\right) = \pm\frac{\pi}{2}$.

(v) Sketch the two loci on the same Argand plane.

(i) $$z_1 = \frac{\sqrt{2}(1-3i)}{2-i} = \frac{\sqrt{2}(1-3i)(2+i)}{(2-i)(2+i)} = \frac{\sqrt{2}(2-5i-3i^2)}{5}$$

$$= \frac{\sqrt{2}(5-5i)}{5} = \sqrt{2}(1-i)$$

$$z_2 = (1+i)^2 = 1+2i+i^2 = 2i$$

(ii) The points P_1 and P_2 are shown in Figure 21(a).

Since $|z_1| = |\sqrt{2}-\sqrt{2}i| = \sqrt{(\sqrt{2})^2+(\sqrt{2})^2} = \sqrt{4} = 2$

and $|z_2| = 2$

it follows that P_1 and P_2 are equidistant from the origin and hence OP_1P_2 is an isosceles triangle.

(iii) If $z = x+iy$ then the locus $|z-z_1| = |z-z_2|$ gives

$$|(x-\sqrt{2})+i(y+\sqrt{2})| = |x+i(y-2)|$$

$$(x-\sqrt{2})^2+(y+\sqrt{2})^2 = x^2+(y-2)^2$$

$$(2\sqrt{2}+4)y = 2\sqrt{2}x \Rightarrow y = \frac{\sqrt{2}x}{2+\sqrt{2}}$$

Rationalizing the denominator $y = \dfrac{\sqrt{2}x(2-\sqrt{2})}{(2+\sqrt{2})(2-\sqrt{2})} = (\sqrt{2}-1)x$

The equation of the locus of P is $y = (\sqrt{2}-1)x$.

(iv) Since $\arg\left(\dfrac{z-z_1}{z-z_2}\right) = \arg(z-z_1)-\arg(z-z_2)$

$$\arg(z-z_1)-\arg(z-z_2) = \pm\tfrac{\pi}{2}$$

This indicates that the angle between the lines joining Q to P_1 and P_2 is 90°. Hence Q moves on a circle with P_1P_2 as diameter.

(v) The sketch is shown in Figure 21(b).

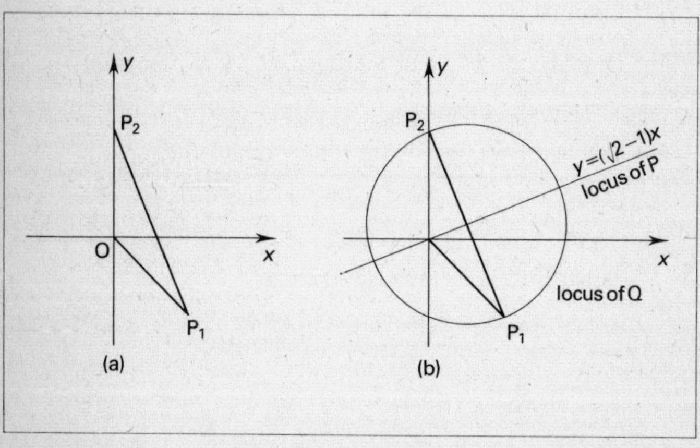

Figure 21

Chapter 10
Differential Equations

Differential equations contain at least one differential coefficient. The study of such equations and their solutions is extensive but syllabuses restrict the range to first order equations using the method of separating the variable. Numerical solutions by the step by step process may also be required.

Multiple choice questions (Type A) Select the correct answer

Example 1

Given that $y = 1$ when $x = 1$ the solution of the equation $x^2 \dfrac{dy}{dx} = y$ is

A $\ln y = \dfrac{1}{x} - 1$ **B** $\ln y = -\dfrac{1}{x}$ **C** $\ln y = 1 - \dfrac{1}{x}$ **D** $\ln y = -\dfrac{2}{x^3}$

E $\ln y = 2 - \dfrac{2}{x}$

Separating the variable

$$x^2 \frac{dy}{dx} = y \Rightarrow \frac{1}{y}\frac{dy}{dx} = \frac{1}{x^2}$$

Integrating with respect to x

$$\int \frac{1}{y}\frac{dy}{dx}\,dx = \int \frac{1}{x^2}\,dx \Rightarrow \int \frac{1}{y}\,dy = \int \frac{1}{x^2}\,dx$$

$\therefore \ \ln y = -\dfrac{1}{x} + c$

$y = 1$ when $x = 1 \Rightarrow c = 1$.

The particular solution is $\ln y = 1 - \dfrac{1}{x}$. ANSWER C

Example 2

The general solution of the equation $y\dfrac{dy}{dx} + x = 1$ defines a family of

A circles, centre (0,0) **B** circles, centre (1,0)
C parabolas, vertex (1,0) **D** parabolas, vertex (0,0)
E ellipses, centre (0,0)

In this equation the variables can be separated to give

$$y\frac{dy}{dx} = 1-x$$

Integrating with respect to x

$$\int y\frac{dy}{dx}\,dx = \int 1-x\,dx \Rightarrow \int y\,dy = \int 1-x\,dx$$

$$\therefore \quad \tfrac{1}{2}y^2 = x-\tfrac{1}{2}x^2+B \Rightarrow y^2+x^2-2x = 2B$$

or $(x-1)^2+y^2 = C$ where $C = 2B+1$.

This general solution represents a family of circles with centres at (1,0).

ANSWER B

Example 3
Which of the following figures could represent part of the family of solution curves of the equation $\frac{dy}{dx} = 5-y$?

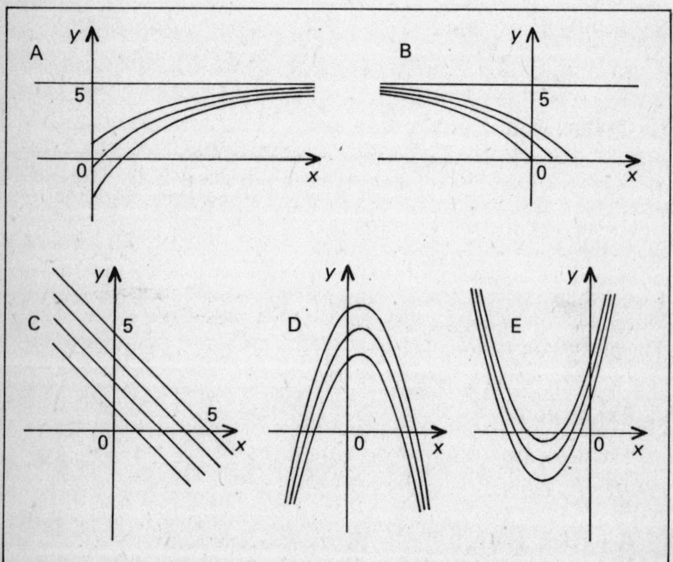

Figure 22

The first order equation can be rewritten by separating the variable.

$$\frac{1}{5-y}\frac{dy}{dx} = 1$$

Integrating with respect to x

$$\int \frac{1}{5-y}\frac{dy}{dx}\,dx = \int 1\,dx \quad \text{or} \quad \int \frac{1}{5-y}\,dy = \int 1\,dx$$

$$\Rightarrow -\ln(5-y) = x - \ln k \quad \text{or} \quad \ln\left(\frac{k}{5-y}\right) = x$$

$$\Rightarrow \frac{k}{5-y} = e^{x} \Rightarrow ke^{-x} = 5-y \Rightarrow y = 5-ke^{-x}$$

As $x \to \infty$, $y \to 5$ and hence whatever boundary conditions are given all the solution curves will be asymptotic to the line $y = 5$. Figure A is the only sketch which shows this property.

ANSWER A

General questions

These examples illustrate the various types of question that may be set including the formulation of a suitable differential equation to describe a model.

> **Example 4**
> In a diluted solution, the rate of decrease of concentration is proportional to the concentration C. If $C = 0.02\,\text{gm/cm}^3$ at time $t = 0$ and $C = 0.01\,\text{gm/cm}^3$ after 2 hours, what will be the concentration after 4 hours?

Since the rate of decrease of the concentration is proportional to the concentration at a given time

$$\frac{dC}{dt} \propto C \Rightarrow \frac{dC}{dt} = kC$$

Separating the variable and integrating with respect to t

$$\int \frac{1}{C}\frac{dC}{dt}\,dt = \int -k\,dt \Rightarrow \int \frac{1}{C}\,dC = \int -k\,dt$$

$$\Rightarrow \ln C = -kt + \ln A \quad \text{or} \quad \ln\left(\frac{C}{A}\right) = -kt$$

Hence the solution is $C = A\,e^{-kt}$.

Now when $t = 0$, $C = 0.02 \Rightarrow A = 0.02$.
Also when $t = 2$, $C = 0.01 \Rightarrow 0.01 = 0.02\,e^{-2k}$
$\Rightarrow e^{-2k} = \frac{1}{2} \Rightarrow -2k = \ln\frac{1}{2} \Rightarrow k = \frac{1}{2}\ln 2$.
Hence $C = 0.02\,e^{-(\frac{1}{2}\ln 2)t}$
$\therefore$ when $t = 4$, $C = 0.02\,e^{-2\ln 2} = 0.02 \times 0.25 = 0.005$.

The concentration after 4 hours is 0.005 gm/cm^3.

Example 5

Solve the differential equation $\dfrac{dy}{dx} + \sec^2 y = 0$.

Rearranging and integrating with respect to x this becomes

$$\int \cos^2 y \frac{dy}{dx}\,dx = -\int dx \Rightarrow \int \cos^2 y\,dy = -\int dx$$

Hence $$\tfrac{1}{2}\int (1+\cos 2y)\,dy = -x+c$$

$\therefore$ $$\frac{y}{2} + \frac{\sin 2y}{4} = -x+c$$

The general solution is $x = \frac{1}{4}(4c - 2y - \sin 2y)$.

Example 6

Solve the differential equation $\dfrac{1}{x}\dfrac{dy}{dx} - y^2 = 1$.

This is an equation where the variables can be separated.

Hence $$\frac{1}{x}\frac{dy}{dx} = y^2+1 \Rightarrow \frac{1}{y^2+1}\frac{dy}{dx} = x$$

Integrating with respect to x

$$\int \frac{1}{y^2+1}\,dy = \int x\,dx \Rightarrow \tan^{-1} y = \tfrac{1}{2}x^2 + c$$

Example 7

Solve the equation $(1+e^y)\dfrac{dy}{dx} = e^{2y}\sin x$ given that $y = 0$ when $x = 0$.

The variables can be separated in this equation and so

$$\left(\frac{1+e^y}{e^{2y}}\right)\frac{dy}{dx} = \sin x \text{ or } (e^{-2y}+e^{-y})\frac{dy}{dx} = \sin x$$

Integrating with respect to x

$$\int e^{-2y}+e^{-y}\,dy = \int \sin x\,dx$$

$$\therefore \quad -\tfrac{1}{2}e^{-2y}-e^{-y} = -\cos x + c$$

The general solution is $\frac{1}{2}e^{-2y}+e^{-y} = \cos x - c$.
Now $y = 0$ when $x = 0 \Rightarrow c = -\frac{1}{2}$.
The particular solution is $\cos x = \frac{1}{2}(e^{-2y}+2e^{-y}-1)$.

Example 8

Solve the equation $x^2\dfrac{dy}{dx} = 4x^2+5xy+y^2$.

Since the equation is of the form $P\dfrac{dy}{dx} = Q$ where P and Q are homogeneous functions of x and y of the same degree we divide by x^2 to give

$$\frac{dy}{dx} = 4+\frac{5y}{x}+\left(\frac{y}{x}\right)^2$$

$$\text{Let } \frac{y}{x} = v \Rightarrow y = vx \Rightarrow \frac{dy}{dx} = v+x\frac{dv}{dx}$$

Substituting, the equation becomes

$$v+x\frac{dv}{dx} = 4+5v+v^2 \quad \text{or} \quad x\frac{dv}{dx} = 4+4v+v^2$$

Hence

$$\frac{1}{(2+v)^2}\frac{dv}{dx} = \frac{1}{x}$$

Integrating with respect to x

$$\int (2+v)^{-2}\,dv = \int \frac{1}{x}\,dx \Rightarrow -(2+v)^{-1} = \ln kx$$

But $v = \dfrac{y}{x}$ and thus

$$-\left(2+\frac{y}{x}\right)^{-1} = \ln kx \quad \text{or} \quad -\frac{x}{(2x+y)} = \ln kx$$

There are many differential equations for which exact solutions cannot be found. In such cases a numerical approximation process can be used which produces values that nearly lie on the solution curve.

A step by step method takes a starting point on the curve and evaluates a neighbouring point by assuming that the curve approximates to the tangent to the curve using the linear approximation $\delta y \simeq f'(x)\delta x$.

A better result can be achieved by using a quadratic approximation, of the form

$$\delta y \simeq f'(x)\delta x + \tfrac{1}{2}f''(x)(\delta x)^2$$

In this case an initial value is required for $f''(x)$ as well.

In Examples 9 and 10, first and second order approximations have been used and give figures which closely match the results that can, in these cases, be achieved by direct integration.

Example 9

Solve the equation $\dfrac{dy}{dx} = \dfrac{1}{x}$ for values of x from 1.0 to 2.0 in steps of 0.2, given that $y = 2$ when $x = 1$.

The working is best done in tabular form. Use the first approximation.

Interval	δx	$f'(x)$	δy	x	y
				1.0	2.000
$1.0 \leqslant x \leqslant 1.2$	0.2	1.0	0.2		
				1.2	2.200
$1.2 \leqslant x \leqslant 1.4$	0.2	0.833	0.167		
				1.4	2.367
$1.4 \leqslant x \leqslant 1.6$	0.2	0.714	0.143		
				1.6	2.510
$1.6 \leqslant x \leqslant 1.8$	0.2	0.625	0.125		
				1.8	2.635
$1.8 \leqslant x \leqslant 2.0$	0.2	0.556	0.111		
				2.0	2.746

Clearly, by integration the solution is $y = \ln x + C$ and under the given initial conditions $C = 2$. The particular solution is

$$y = \ln x + 2$$

Chapter 11
Vectors

Questions involving vectors are many and varied, ranging from short questions on the algebra of vectors to geometrical applications and the uses of vector methods in applied mathematics. Different examination boards use different notations and it is wise to be thoroughly conversant with the notation used in each examination paper.

Multiple choice questions (Type A) Select the correct answer

> **Example 1**
> P divides AB in the ratio 1:2. Referred to the origin O, $\mathbf{OP} =$
>
> A $\mathbf{OA}+2\mathbf{OB}$ B $2\mathbf{OA}+\mathbf{OB}$ C $\frac{2}{3}\mathbf{OA}+\frac{1}{3}\mathbf{OB}$
> D $\frac{1}{3}\mathbf{OA}+\frac{2}{3}\mathbf{OB}$ E $2\mathbf{OA}-\mathbf{OB}$

$$\mathbf{OP} = \mathbf{OA}+\mathbf{AP} = \mathbf{OA}+\tfrac{1}{3}\mathbf{AB}$$

$$= \mathbf{OA}+\tfrac{1}{3}(\mathbf{AO}+\mathbf{OB})$$

$$= \mathbf{OA}-\tfrac{1}{3}\mathbf{OA}+\tfrac{1}{3}\mathbf{OB}$$

$$= \tfrac{2}{3}\mathbf{OA}+\tfrac{1}{3}\mathbf{OB}$$

ANSWER C

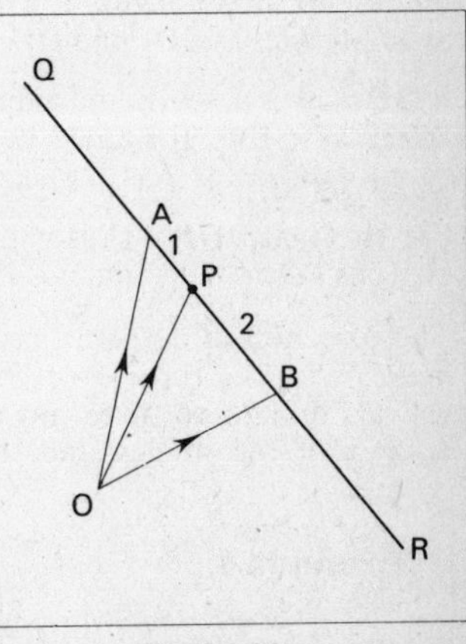

Figure 23

This is the ratio theorem which can be quoted. If the ratio is 1:2 the fractions are $\frac{1}{3}$ and $\frac{2}{3}$. To make sure these are the right way round ($\frac{2}{3}$ with **OA**, $\frac{1}{3}$ with **OB**) remember that P is nearer A and will contain more **OA** than **OB**.

> **Example 2**
> Q divides AB in the ratio $-1:2$. $\mathbf{OQ} =$
>
> A $-\mathbf{OA}+2\mathbf{OB}$ B $\mathbf{OA}-2\mathbf{OB}$ C $\frac{1}{3}\mathbf{OA}-\frac{2}{3}\mathbf{OB}$
> D $\frac{2}{3}\mathbf{OA}-\frac{1}{3}\mathbf{OB}$ E $2\mathbf{OA}-\mathbf{OB}$

With a negative ratio, Q is either on AB produced or BA produced. $AQ:QB = -1:2 \Rightarrow$ the distance AQ compared with QB is 1:2 in magnitude and opposite in direction. This can happen only when Q is outside A (on BA produced) as shown in Figure 23. It is easier to regard A as the mid-point of Q and B and apply the ratio theorem to get

$$\mathbf{OA} = \tfrac{1}{2}\mathbf{OQ} + \tfrac{1}{2}\mathbf{OB} \Rightarrow 2\mathbf{OA} = \mathbf{OQ} + \mathbf{OB} \Rightarrow \mathbf{OQ} = 2\mathbf{OA} - \mathbf{OB}$$

ANSWER E

If R divides AB in the ratio $-2:1$ (or $2:-1$) $AR:RB = -2:1$ and R is outside B. In this case $\mathbf{OR} = 2\mathbf{OB} - \mathbf{OA}$ and as R is nearer to B than A there is more of **OB** than **OA** in the composition of **OR**.

If P (or Q) is anywhere on AB, the ratios add up to 1.

Example 3
The direction cosines of the vector $3\mathbf{i} - 4\mathbf{j} + 12\mathbf{k}$ are

A $3, -4, 12$ **B** $3, 4, 12$ **C** $\frac{3}{19}, -\frac{4}{19}, \frac{12}{19}$ **D** $\frac{3}{11}, -\frac{4}{11}, \frac{12}{11}$
E $\frac{3}{13}, -\frac{4}{13}, \frac{12}{13}$

The direction cosines of a vector are the cosines of the 3 angles that the vector makes with the positive directions of the three co-ordinate axes Ox, Oy and Oz.

If $\mathbf{OP} = 3\mathbf{i} - 4\mathbf{j} + 12\mathbf{k}$ and a line is drawn from P to meet Ox at right angles at x, then the angle POx is given by $\cos POx = \frac{3}{13}$ since the length OP is $\sqrt{3^2 + 4^2 + 12^2} = 13$.

For the angles OP makes with Oy and Oz a similar result holds to give $\cos POy = -\frac{4}{13}$ and $\cos POz = \frac{12}{13}$. ANSWER E

The direction of a vector can be specified by the angles which it makes with the three coordinate axes. This method is the easiest way to find these three angles and since it involves finding the cosines of the angles, the three ratios are called the direction cosines.

Example 4
The plane $x - 2y + 3z = 6$ meets the coordinate axes Ox, Oy and Oz

A at the origin
B at (1,0,0) (0,−2,0) (0,0,3)
C at (3,0,0) (0,2,0) (0,0,1)
D at (6,0,0) (0,3,0) (0,0,2)
E at (6,0,0) (0,−3,0) (0,0,2)

If the plane meets Ox at X then X is given by $y=0$ and $z=0 \Rightarrow x=6$.
Similarly $-2y=6 \Rightarrow y=-3$ at $Y(0,-3,0)$ and Z is $(0,0,2)$.
E is the correct answer. ANSWER E

Example 5
The scalar product of the vectors $\mathbf{i}+2\mathbf{j}+3\mathbf{k}$ and $3\mathbf{i}-2\mathbf{j}+\mathbf{k}$ is

A $3\mathbf{i}+4\mathbf{j}+3\mathbf{k}$ **B** $3\mathbf{i}-4\mathbf{j}+3\mathbf{k}$ **C** 10 **D** 2 **E** 36

The scalar product of two vectors is a number (a scalar is a number). The respective components are multiplied together and then added.

$$\mathbf{a}\times\mathbf{b} = (\mathbf{i}+2\mathbf{j}+3\mathbf{k})\times(3\mathbf{i}-2\mathbf{j}+\mathbf{k}) = 3-4+3 = 2$$

ANSWER D

Example 6
The vector product of the vectors $\mathbf{a}=\mathbf{i}+2\mathbf{j}+3\mathbf{k}$ and $\mathbf{b}=3\mathbf{i}-2\mathbf{j}+\mathbf{k}$ is

A $3\mathbf{i}+4\mathbf{j}+3\mathbf{k}$ **B** $3\mathbf{i}-4\mathbf{j}+3\mathbf{k}$ **C** $8\mathbf{i}-8\mathbf{j}-8\mathbf{k}$
D $8\mathbf{i}+8\mathbf{j}-8\mathbf{k}$ **E** $\mathbf{i}+\mathbf{j}-\mathbf{k}$

$$\mathbf{a}\times\mathbf{b} = \begin{vmatrix} \mathbf{i} & \mathbf{j} & \mathbf{k} \\ 1 & 2 & 3 \\ 3 & -2 & 1 \end{vmatrix} = \mathbf{i}(+2+6)-\mathbf{j}(1-9)+\mathbf{k}(-2-6)$$

$$= 8\mathbf{i}+8\mathbf{j}-8\mathbf{k}$$

ANSWER D

The usual mistake is to get the sign of the middle component ($\mathbf{j}$) wrong. As a check $\mathbf{a}\times\mathbf{b}$ is perpendicular to both $\mathbf{a}$ and $\mathbf{b}$.

Multiple choice questions (Type B) Answer according to the table

A	B	C	D	E
1, 2, 3 correct	1, 3 only	2, 3 only	2 only	3 only

Example 7
The vector equation of AB is

1 $\mathbf{r}=\mathbf{OA}+t\mathbf{AB}$ **2** $\mathbf{r}=\mathbf{OA}+s\mathbf{BA}$ **3** $\mathbf{r}=\mathbf{OB}+p\mathbf{BA}$

The vector equation of a straight line is not unique: there are several possible forms. Three are given in this question. If the equation is $\mathbf{r} = \mathbf{OC} + t\mathbf{EF}$, C must be a point on the line and the vector $\mathbf{EF}$ must represent the direction of the line. All three answers are correct. ANSWER A

Example 8
In Figure 24, $\mathbf{OD}$ =

1 $\frac{1}{3}\mathbf{OA}+\frac{1}{3}\mathbf{OB}+\frac{1}{3}\mathbf{OC}$ **2** $\frac{5}{9}\mathbf{OA}+\frac{4}{9}\mathbf{OC}+\frac{1}{3}\mathbf{OB}$
3 $\frac{2}{3}(\frac{1}{2}\mathbf{OA}+\frac{1}{2}\mathbf{OB})+\frac{1}{3}\mathbf{OC}$

By coordinates

$$\mathbf{OA}+\mathbf{OB}+\mathbf{OC} = 6\mathbf{j}+6\mathbf{i}+9\mathbf{i}+6\mathbf{j} = 15\mathbf{i}+12\mathbf{j}$$

$$\tfrac{1}{3}(\mathbf{OA}+\mathbf{OB}+\mathbf{OC}) = 5\mathbf{i}+4\mathbf{j}$$

Statement 1 is correct.

$$\tfrac{5}{9}\mathbf{OA}+\tfrac{4}{9}\mathbf{OC}+\tfrac{1}{3}\mathbf{OB} = 3\tfrac{1}{3}\mathbf{j}+4\mathbf{i} + 2\tfrac{2}{3}\mathbf{j}+2\mathbf{i} = 6\mathbf{i}+6\mathbf{j}$$

Statement 2 is not correct.

Statement 3 is the same as statement 1, but suggests another method.

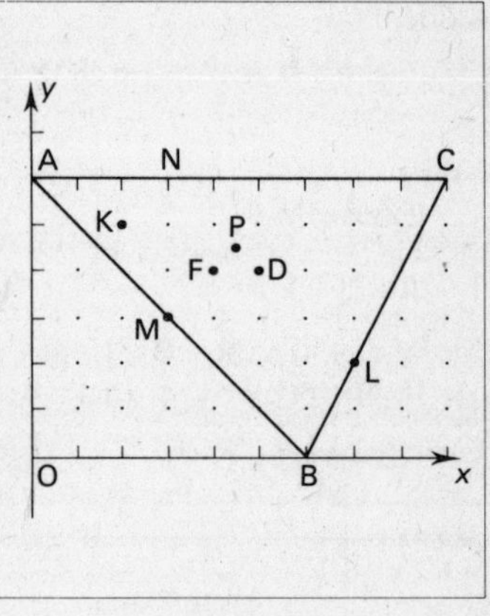

Figure 24

$\mathbf{OM} = \frac{1}{2}\mathbf{OA}+\frac{1}{2}\mathbf{OB}$ and D divides CM in the ratio 2:1 so

$$\mathbf{OD} = \tfrac{1}{3}\mathbf{OC}+\tfrac{2}{3}\mathbf{OM} = \tfrac{2}{3}(\tfrac{1}{2}\mathbf{OA}+\tfrac{1}{2}\mathbf{OB})+\tfrac{1}{3}\mathbf{OC} = \tfrac{1}{3}\mathbf{OA}+\tfrac{1}{3}\mathbf{OB}+\tfrac{1}{3}\mathbf{OC}$$

ANSWER B

Knowing the ratios, there is no need for coordinates; statements 1 and 3 are correct. D is the centroid (centre of mass or gravity) of the triangle.

Example 9
In Figure 24 $\mathbf{OF}$ in terms of $\mathbf{OA}$, $\mathbf{OB}$ and $\mathbf{OC}$ is equal to

1 $\frac{4}{9}\mathbf{OA}+\frac{2}{9}\mathbf{OB}+\frac{1}{3}\mathbf{OC}$ **2** $\frac{2}{9}\mathbf{OA}+\frac{4}{9}\mathbf{OB}+\frac{1}{3}\mathbf{OC}$
3 $\frac{2}{9}\mathbf{OA}+\frac{1}{3}\mathbf{OB}+\frac{4}{9}\mathbf{OC}$

All three answers are different so only one can be right.
F divides BN in the ratio $2:1 \Rightarrow \mathbf{OF} = \frac{2}{3}\mathbf{ON} + \frac{1}{3}\mathbf{OB}$
N divides AC in the ratio $1:2 \Rightarrow \mathbf{ON} = \frac{1}{3}\mathbf{OA} + \frac{2}{3}\mathbf{OC}$

$$\mathbf{OF} = \tfrac{2}{3}(\tfrac{1}{3}\mathbf{OA} + \tfrac{2}{3}\mathbf{OC}) + \tfrac{1}{3}\mathbf{OB} = \tfrac{2}{9}\mathbf{OA} + \tfrac{1}{3}\mathbf{OB} + \tfrac{4}{9}\mathbf{OC}$$

Notice that the coefficients add up to 1. Statement 3 only is correct. ANSWER E

Example 10
The length of the vector $3\mathbf{i} + 4\mathbf{j} + 12\mathbf{k}$ is

1 $3+4+12$ **2** 13 **3** $\sqrt{3^2+4^2+12^2}$

To find the length of a vector in three dimensions we use a three-dimensional form of Pythagoras' theorem as in answer 3.

Statements 3 and 2 give the same value 13 and are both correct. As a simple check, remember that a rule for vectors must be true in two dimensions as well as in three, and discounting the third component ($12\mathbf{k}$) we know that the length of $3\mathbf{i} + 4\mathbf{j}$ is 5 (3,4,5 triangle) so statement 1 ($3+4+12$) cannot possibly be correct. ANSWER C

Example 11
The angle θ between the vectors $\mathbf{a}$ and $\mathbf{b}$ is given by

1 $\cos\theta = \dfrac{\mathbf{a}\times\mathbf{b}}{ab}$ **2** $\tan\theta = \dfrac{|\mathbf{a}+\mathbf{b}|}{ab}$ **3** $\sin\theta = \dfrac{|\mathbf{a}\times\mathbf{b}|}{ab}$

By definition $\mathbf{a}\times\mathbf{b} = ab\cos\theta$ where $a = |\mathbf{a}|$ and $b = |\mathbf{b}|$. $\mathbf{a}\times\mathbf{b}$ can be calculated from components and θ determined so statement 1 is correct. Statement 2 is incorrect: if $\mathbf{a} = \mathbf{i}$ and $\mathbf{b} = \mathbf{j}$ then $|\mathbf{a}+\mathbf{b}| = |\mathbf{i}+\mathbf{j}| = \sqrt{2}$. Statement 2 gives $\tan\theta = \dfrac{\sqrt{2}}{1} \Rightarrow \theta = \tan^{-1}\sqrt{2} = 54.7°$, but $\mathbf{i}$ and $\mathbf{j}$ are perpendicular.

The vector product $\mathbf{a}\times\mathbf{b} = ab\sin\theta\,\mathbf{n}$ where $\mathbf{n}$ is a unit vector perpendicular to $\mathbf{a}$ and $\mathbf{b}$. $\mathbf{a}\times\mathbf{b}$ can be calculated from components and $\sin\theta$ then calculated. Statement 3 is correct. ANSWER B

These methods for finding the angle between vectors are used in Example 18.

> **Example 12**
> The angle between the diagonals of a cube is
>
> **1** $90°$ **2** $60°$ **3** $\cos^{-1}\frac{1}{3}$

There are 8 diagonals in a cube, two of which, AOC and BOD, are shown in Figure 25. With O as origin and $A(1,1,1)$ and $B(1,1,-1)$, $AB=2$ and $AD=2\sqrt{2}$.

The angle between the diagonals is either $\angle BOA$ or $\angle AOD$ which are different but add up to 180°. In $\triangle OAM$

$\cos\theta = OM/AO = \sqrt{2}/\sqrt{3}$

$\cos 2\theta = 2\cos^2\theta - 1 = 1\frac{1}{3} - 1 = \frac{1}{3}$

$\angle BOA = 2\theta = \cos^{-1}\frac{1}{3} = 70.5°$

Statement 3 only is correct.

ANSWER E

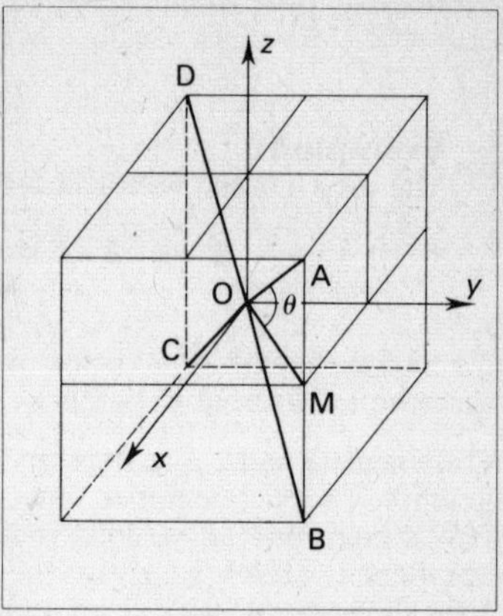

Figure 25

General questions

> **Example 13**
> If D is the centroid of $\triangle ABC$, show that $\mathbf{DA}+\mathbf{DB}+\mathbf{DC}=\mathbf{0}$. Use Figure 24 for reference and check the result using the coordinates to specify the vectors.

Referring to Figure 24 M is the mid-point of $\mathbf{AB} \Rightarrow \mathbf{DM} = \frac{1}{2}\mathbf{DA}+\frac{1}{2}\mathbf{DB}$

$\mathbf{DA}+\mathbf{DB} = 2\mathbf{DM} = \mathbf{CD} = -\mathbf{DC} \Rightarrow \mathbf{DA}+\mathbf{DB}+\mathbf{DC} = 0$

$\mathbf{DA}+\mathbf{DB}+\mathbf{DC} = (-5\mathbf{i}+2\mathbf{j})+(\mathbf{i}-4\mathbf{j})+(4\mathbf{i}+2\mathbf{j}) = 0$

> **Example 14**
> Find the centroid of the tetrahedron $A(0,6,0)$, $B(6,0,0)$, $C(9,6,0)$, $V(5,4,8)$.

The centroid G is the intersection of lines joining each vertex to the centroid of the opposite face. The centroid of $\triangle ABC$ has already been calculated in two dimensions in Example 8 and is denoted by $D(5,4,0)$ in Figure 24.

G divides VD in the ratio 3:1 so $\mathbf{OG} = \frac{1}{4}\mathbf{OV} + \frac{3}{4}\mathbf{OD}$

$\mathbf{OG} = \frac{1}{4}(5\mathbf{i}+4\mathbf{j}+8\mathbf{k}) + \frac{3}{4}(5\mathbf{i}+4\mathbf{j}) = \frac{1}{4}(20\mathbf{i}+16\mathbf{j}+8\mathbf{k}) = 5\mathbf{i}+4\mathbf{j}+2\mathbf{k}$

As a check, G should divide AE (E is the centroid of BCV) in the ratio 3:1 so that

$$\begin{aligned}\mathbf{OG} = \tfrac{1}{4}\mathbf{OA} + \tfrac{3}{4}\mathbf{OE} &= \tfrac{1}{4}(6\mathbf{j}) + \tfrac{3}{4}\cdot\tfrac{1}{3}(\mathbf{OB}+\mathbf{OC}+\mathbf{OV}) \\ &= \tfrac{1}{4}(6\mathbf{j}) + \tfrac{1}{4}(6\mathbf{i}+9\mathbf{i}+6\mathbf{j}+5\mathbf{i}+4\mathbf{j}+8\mathbf{k}) \\ &= \tfrac{1}{4}(20\mathbf{i}+16\mathbf{j}+8\mathbf{k}) \\ &= 5\mathbf{i}+4\mathbf{j}+2\mathbf{k}\end{aligned}$$

From this we can see that
$$\begin{aligned}\mathbf{OG} &= \tfrac{1}{4}(\mathbf{OA}+\mathbf{OB}+\mathbf{OC}+\mathbf{OV}) \\ &= \tfrac{1}{2}(\tfrac{1}{2}\mathbf{OA}+\tfrac{1}{2}\mathbf{OB}) + \tfrac{1}{2}(\tfrac{1}{2}\mathbf{OC}+\tfrac{1}{2}\mathbf{OV}) \\ &= \tfrac{1}{2}\mathbf{OM} + \tfrac{1}{2}\mathbf{OW}\end{aligned}$$

M is the mid-point of AB and W is the mid-point of CV.
G is the centre of the line joining the mid-points of opposite edges.

Example 15
Find the equation of the plane through $A(3,2,1)$, $B(1,2,3)$, $C(6,6,1)$ and deduce the points X, Y and Z where the plane meets the axes.

A plane has a vector equation and a Cartesian equation.

Method 1 The Cartesian equation is given by $ax+by+cz=d$
$A(3,2,1)$ lies on the plane $\Rightarrow 3a+2b+c=d$ (1)
$B(1,2,3)$ lies on the plane $\Rightarrow a+2b+3c=d$ (2)
$C(6,6,1)$ lies on the plane $\Rightarrow 6a+6b+c=d$ (3)

(1) − (2) gives $2a-2c=0 \Rightarrow a=c$

With $a=c$ (2) becomes $4a+2b=d$
(3) becomes $7a+6b=d$ $\quad 5a=2d \Rightarrow a=c=\frac{2}{5}d$

$2b = d-4a = d-\frac{8}{5}d = -\frac{3}{5}d \Rightarrow b = -\frac{3}{10}d$

$ax+by+cz=d$ becomes $\frac{2}{5}dx - \frac{3}{10}dy + \frac{2}{5}dz = d \Rightarrow 4x-3y+4z=10$.
The plane meets the x-axis where $y=0$ and $z=0 \Rightarrow x=2\frac{1}{2}$ and X is $(2\frac{1}{2},0,0)$, Y is $(0,-3\frac{1}{3},0)$ and Z is $(0,0,2\frac{1}{2})$.

Method 2 The vector equation of the plane is given by $\mathbf{r} = x\mathbf{i}+y\mathbf{j}+z\mathbf{k} = \mathbf{OA}+s\mathbf{AB}+t\mathbf{AC}$

$$\begin{aligned}&\Rightarrow \mathbf{r} = \mathbf{OA}+s(\mathbf{OB}-\mathbf{OA})+t(\mathbf{OC}-\mathbf{OA}) \\ &\Rightarrow \mathbf{r} = 3\mathbf{i}+2\mathbf{j}+\mathbf{k}+s(-2\mathbf{i}+2\mathbf{k})+t(3\mathbf{i}+4\mathbf{j})\end{aligned}$$

Different values of s and t give different points on the plane, e.g. $s=t=0 \Rightarrow \mathbf{r}=\mathbf{OA}$; $s=1, t=0 \Rightarrow \mathbf{r}=\mathbf{OB}=\mathbf{i}+2\mathbf{j}+\mathbf{k}$.

To find the Cartesian equation s and t must be eliminated.

Equating coefficients of

$$\mathbf{i} \Rightarrow x = 3-2s+3t \qquad (1)$$
$$\mathbf{j} \Rightarrow y = 2+4t \Rightarrow t = \tfrac{1}{4}(y-2)$$
$$\mathbf{k} \Rightarrow z = 1+2s \Rightarrow s = \tfrac{1}{2}(z-1)$$

Substituting for s and t in (1) $\Rightarrow$ $x = 3-(z-1)+\tfrac{3}{4}(y-2)$

$$\Rightarrow 4x = 12-4z+4+3y-6$$
$$\Rightarrow 4x-3y+4z = 10$$

X, Y and Z are found as in Method 1.

Method 3 For the plane $ax+by+cz = d$ the vector $a\mathbf{i}+b\mathbf{j}+c\mathbf{k}$ is the normal vector, i.e., it is a vector which is perpendicular to the plane and in effect gives the 'direction' of the plane. This normal vector is perpendicular to all vectors in the plane. To find it, form the vector product of two vectors in the plane (in different directions), e.g., **AB** and **AC**.

$$\mathbf{AB}\times\mathbf{AC} = (-2\mathbf{i}+2\mathbf{k})\times(3\mathbf{i}+4\mathbf{j}) = \begin{vmatrix} \mathbf{i} & \mathbf{j} & \mathbf{k} \\ -2 & 0 & 2 \\ 3 & 4 & 0 \end{vmatrix} = -8\mathbf{i}+6\mathbf{j}-8\mathbf{k}$$

The plane ABC has the form $-8x+6y-8z = d$ and since $A(1,2,3)$ lies in the plane $-8+12-24 = d \Rightarrow d = -20$.
The plane has equation $-8x+6y-8z = -20 \Rightarrow 4x-3y+4z = 10$.

Method 3 involves working out 3 second order determinants rather than 3 simultaneous equations and is therefore the quickest method.

Example 16
Find the distance of the point $D(3,4,5)$ from the plane through $A(1,2,3)$, $B(3,2,1)$ and $C(6,6,1)$.

Plane ABC was found in the previous example to be $4x-3y+4z = 10$. The distance of the point (x_1,y_1,z_1) from the plane $ax+by+cz = d$ is given by

$$\frac{ax_1+by_1+cz_1-d}{\sqrt{(a^2+b^2+c^2)}} = \frac{12-12+20-10}{\sqrt{(4^2+(-3)^2+4^2)}} = \frac{10}{\sqrt{41}} \simeq 1.56.$$

The distance from the origin O to the plane ABC is $-10/\sqrt{41} \simeq -1.56$, the negative sign indicating that O is on the opposite side of the plane to D.

Example 17
With reference to Figure 24 find the coordinates of P, the circumcentre of ABC and check that the length of vectors $\mathbf{PA} = \mathbf{a}$, $\mathbf{PB} = \mathbf{b}$ and $\mathbf{PC} = \mathbf{c}$ are equal.
(b) Defining $\mathbf{PH} = \mathbf{h} = \mathbf{a} + \mathbf{b} + \mathbf{c}$ show that $\mathbf{CH}$ is perpendicular to $\mathbf{AB}$ and $\mathbf{BH}$ is perpendicular to $\mathbf{AC}$. Use this to find H.
(c) How does H relate to $\triangle ABC$ and what is the relation between P, D and H.

(a) The circumcentre of the triangle is the intersection of the mediators (perpendicular bisectors) of the sides.
The mediator of AB is MF whose equation is $y = x$.
The mediator of AC is $x = 4\frac{1}{2}$.
These lines intersect at $P(4\frac{1}{2}, 4\frac{1}{2})$ so P is the circumcentre.

(b) $\mathbf{PA} = -4\frac{1}{2}\mathbf{i} + 1\frac{1}{2}\mathbf{j}$; $\mathbf{PB} = 1\frac{1}{2}\mathbf{i} - 4\frac{1}{2}\mathbf{j}$; $\mathbf{PC} = 4\frac{1}{2}\mathbf{i} + 1\frac{1}{2}\mathbf{j}$
These three vectors have length $\sqrt{(1\frac{1}{2}^2 + 4\frac{1}{2}^2)} = \frac{1}{2}\sqrt{(9+81)} = \frac{1}{2}\sqrt{90} = 1\frac{1}{2}\sqrt{10}$. Therefore, **a**, **b** and **c** have the same length.

$\mathbf{h} = \mathbf{a} + \mathbf{b} + \mathbf{c} \Rightarrow \mathbf{h} - \mathbf{c} = \mathbf{a} + \mathbf{b}$
which is perpendicular to $\mathbf{a} - \mathbf{b}$ since **a** and **b** are equal in length.

The parallelogram $OACB$ where $\mathbf{OA} = \mathbf{a}$, $\mathbf{OB} = \mathbf{b}$, $\mathbf{OC} = \mathbf{a} + \mathbf{b}$ has diagonals OC and AB represented by $\mathbf{OC} = \mathbf{a} + \mathbf{b}$ and $\mathbf{BA} = \mathbf{a} - \mathbf{b}$.
If a and b are equal in length, $OACB$ is a rhombus and the diagonals are perpendicular.

a and **b** have the same length $\Leftrightarrow \mathbf{a} + \mathbf{b}$ and $\mathbf{a} - \mathbf{b}$ are perpendicular.
$\mathbf{h} - \mathbf{c} = \mathbf{BH}$ and $\mathbf{a} - \mathbf{b} = \mathbf{BA} \Rightarrow CH$ and AB are perpendicular.
Similarly, $\mathbf{h} - \mathbf{b} = \mathbf{a} + \mathbf{c}$ which is perpendicular to $\mathbf{a} - \mathbf{c}$.
$\mathbf{h} - \mathbf{b} = \mathbf{BH}$ and $\mathbf{a} - \mathbf{c} = \mathbf{CA} \Rightarrow BH$ and AC are perpendicular.
CH and BH are altitudes of the $\triangle ABC$ which intersect at H with coordinates (6,3).
As a check $\mathbf{h} = \mathbf{a} + \mathbf{b} + \mathbf{c} = \mathbf{PA} + \mathbf{PB} + \mathbf{PC} = 1\frac{1}{2}\mathbf{i} - 1\frac{1}{2}\mathbf{j} = \mathbf{PH}$

$\mathbf{OH} = \mathbf{OP} + \mathbf{PH} = 4\frac{1}{2}\mathbf{i} + 4\frac{1}{2}\mathbf{j} + 1\frac{1}{2}\mathbf{i} - 1\frac{1}{2}\mathbf{j} = 6\mathbf{i} + 3\mathbf{j}$ as above.

(c) H is the orthocentre of $\triangle ABC$ (intersection of the altitudes). By a similar argument to those above, HA and BC are perpendicular. P(circumcentre), D(centroid) and H(orthocentre) are colinear and D divides PH in the ratio 1:2.

Example 18
Find the angle between the vectors $\mathbf{a} = \mathbf{i} + 2\mathbf{j} + 3\mathbf{k}$ and $\mathbf{b} = 3\mathbf{i} - 2\mathbf{j} + \mathbf{k}$ by using (i) the scalar product and (ii) the vector product.

(i) $\mathbf{a} \times \mathbf{b} = (\mathbf{i}+2\mathbf{j}+3\mathbf{k}) \times (2\mathbf{i}-2\mathbf{j}+\mathbf{k}) = 3-4+3 = 2$

$\mathbf{a} \times \mathbf{b} = ab\cos\theta$ where $a = |\mathbf{a}| = \sqrt{1^2+2^2+3^2} = \sqrt{14}$ and $b = |\mathbf{b}| = \sqrt{14} \Rightarrow \sqrt{14}\sqrt{14}\cos\theta = 2 \Rightarrow \cos\theta = \frac{1}{7} \Rightarrow \theta = 81.8°$

(ii)

$$\mathbf{a} \times \mathbf{b} = \begin{vmatrix} \mathbf{i} & \mathbf{j} & \mathbf{k} \\ 1 & 2 & 3 \\ 3 & -2 & 1 \end{vmatrix} = 8\mathbf{i}+8\mathbf{j}-8\mathbf{k}$$

$$\Rightarrow |\mathbf{a} \times \mathbf{b}| = \sqrt{8^2+8^2+(-8)^2} = 8\sqrt{3}$$

$|\mathbf{a} \times \mathbf{b}| = ab\sin\theta$ where $a = b = \sqrt{14} \Rightarrow \sqrt{14}\sqrt{14}\sin\theta = 8\sqrt{3}$

$$\Rightarrow \sin\theta = 4\sqrt{3}/7 \Rightarrow \theta = 81.8°$$

As a check $\sin^2\theta + \cos^2\theta = \dfrac{16 \times 3}{49} + \dfrac{1}{49} = \dfrac{49}{49} = 1$

> **Example 19**
> Find the line of intersection of the planes $x+2y+3z = 14$ and $2x+5y-z = 9$ and find where this line intersects the plane $3x+8y-4z = 7$.

The equations $\dfrac{x-1}{3} = \dfrac{y-2}{4} = \dfrac{z-3}{5} = t$ or $x = 1+3t$, $y = 2+4t$, $z = 3+5t$ represent a straight line through the point (1,2,3) in the direction $3\mathbf{i}+4\mathbf{j}+5\mathbf{k}$.

Solving the equations

$$x+2y+3z = 14 \qquad (1)$$
$$2x+5y-z = 9 \qquad (2)$$

in that form

$(1)+3\times(2) \Rightarrow 7x+17y = 41$ or $7x-41 = -17y$
$(2)-2\times(1) \Rightarrow y-7z = -19$ or $y = 7z-19$

The line of intersection is

$$\frac{x-\frac{41}{7}}{-17} = \frac{y}{7} = \frac{z-\frac{19}{7}}{1} = t \qquad \text{or} \qquad \left.\begin{aligned} x &= -17t+\tfrac{41}{7} \\ y &= 7t \\ z &= t+\tfrac{19}{7} \end{aligned}\right\} \quad A$$

To find where this line meets the plane $3x+8y-4z = 7$ we find the value of t which allows equations A to satisfy $3x+8y-4z = 7$.

$\Rightarrow 3(-17t+\frac{41}{7})+56t-4(t+\frac{19}{7}) = 7$

$\Rightarrow -51t+52t = 7+\frac{76}{7}-\frac{123}{7} \Rightarrow t = \dfrac{49+76-123}{7} = \frac{2}{7}$

$t = \frac{2}{7} \Rightarrow x = \frac{41}{7} - \frac{34}{7} = 1;\ y = 2;\ z = \frac{21}{7} = 3.$
The three planes intersect in the single point (1,2,3).

If the line of intersection of the first two planes turns out to be parallel to the third plane, there will be no solution to the equation. This will be true if the direction vector of the line of intersection is perpendicular to the normal vector of the third plane. If the line of intersection lies completely in the third plane, the t equation will be satisfied by all values of t.

Chapter 12
Probability

Multiple choice questions (Type A) Select the correct answer

Example 1
The probability of throwing 3 consecutive heads with 3 throws of a coin is

A $\frac{1}{2}+\frac{1}{2}+\frac{1}{2}$ **B** $\frac{1}{6}$ **C** $\frac{1}{2}\times\frac{1}{2}\times\frac{1}{2}$ **D** $\frac{1}{2}$ **E** $\frac{1}{4}$

There are 4 different results with 3 throws of a coin: 3 heads; 3 tails; 2 heads and a tail; 2 tails and a head; but each of the last two can occur in 3 ways making 8 possible outcomes in all. The result, 3 heads, is only one result from these 8:

HHH, HHT, HTH, THH, HTT, THT, TTH, TTT

The probability is one in eight. ANSWER C

Example 2
A set T of numbers has mean 5 and standard deviation 3. If each number in T is doubled, T has mean m and standard deviation s where

A $m = 5, s = 3$ **B** $m = 5, s = 6$ **C** $m = 10, s = 3$
D $m = 10, s = 6$ **E** $m = 10, s = 9$

If $T = x_1, x_2, \ldots, x_n$ then $m = \dfrac{\Sigma x}{n}$ and $s^2 = \dfrac{1}{n}\Sigma(x-m)^2$

For the doubled set $m = \dfrac{\Sigma 2x}{n} = \dfrac{2\Sigma x}{n} = 2\times 5 = 10$ and

$$s^2 = \frac{1}{n}\Sigma(2x-2m)^2 = \frac{1}{n}\Sigma 2^2(x-m)^2 = \frac{4}{n}\Sigma(x-m)^2 = 4\times 9 = 36$$

$s = 6$ and $m = 10$. ANSWER D

Example 3
The probability of drawing an ace from a pack of cards is $\frac{1}{13}$ and the probability of drawing a heart is $\frac{1}{4}$. The probability of drawing the ace of hearts is

A $\frac{1}{4}+\frac{1}{13}$ **B** $\frac{1}{4}\times\frac{1}{13}$ **C** $\frac{1}{4}-\frac{1}{13}$ **D** $\frac{1}{4}\div\frac{1}{13}$ **E** $\frac{4}{13}$

The ace of hearts is one card out of a pack of 52, so the probability of drawing it is $\frac{1}{52}$. ANSWER B

In this case we need the probability of both an ace and a heart, i.e., both events happening:

Probability of A and $B = p(A) \times p(B)$ if the events are independent.

In this case it is easy to see that the other answers are mostly ridiculous.

Example 4
Three discs (identical except for colour) are drawn from a bag containing 2 red, 3 blue and 4 green discs, without replacement. The probability of drawing 1 red and 2 blue discs in that order is

A $\frac{2}{9} \times \frac{3}{8} \times \frac{3}{7}$ **B** $\frac{2}{9} + \frac{3}{8} + \frac{2}{7}$ **C** $\frac{2}{9} \times \frac{3}{8} \times \frac{2}{7}$ **D** $\frac{1}{2} \times \frac{1}{3} \times \frac{1}{3}$ **E** $\frac{2}{5} \times \frac{3}{4} \times \frac{2}{3}$

The probability that the first disc drawn is red is $\frac{2}{9}$.

This leaves 8 discs, 3 of which are blue, so the probability of the second being blue is $\frac{3}{8}$ which leaves 2 blue out of 7 and the probability of the third being blue is $\frac{2}{7}$.
The probability of all three events is $\frac{2}{9} \times \frac{3}{8} \times \frac{2}{7}$. ANSWER C

Example 5
In my tutor group of 15, 5 study Mathematics, 10 English and 2 study both. If I pick one at random the probability that he or she studies neither Mathematics nor English is

A 0 **B** $1 - \frac{1}{3} \times \frac{2}{3}$ **C** $\frac{2}{3} \times \frac{1}{3}$ **D** $\frac{2}{15}$ **E** none of these

Out of 15, 3 study Mathematics and not English and 8 study English and not Mathematics. With 2 studying both, this accounts for 13, so the 2 left study neither, i.e., 2 out of 15. ANSWER D

Example 6
The probability that it will rain on any particular day is $\frac{3}{5}$. The probability that it will rain on 3 out of the next 5 days is

A 1 **B** $3 \times \frac{3}{5}$ **C** $\frac{3}{5} \times \frac{3}{5} \times \frac{3}{5}$ **D** $(\frac{3}{5})^3 \times (\frac{2}{5})^2$ **E** $10 \times (\frac{3}{5})^3 \times (\frac{2}{5})^2$

The question implies that it rains on three days and does not rain on the other two. The probabilities are therefore $\frac{3}{5} \times \frac{3}{5} \times \frac{3}{5} \times \frac{2}{5} \times \frac{2}{5}$ for three days rain and two without but these can be arranged in ${}_5C_3$ ways. Total probability is ${}_5C_3 \times (\frac{3}{5})^3 \times (\frac{2}{5})^2 = 10 \times (\frac{3}{5})^3 \times (\frac{2}{5})^2$.
ANSWER E

Example 7
Ten coins are thrown. The probability of 5 heads appearing is

A $\frac{1}{2}$ **B** $(\frac{1}{2})^5 \times (\frac{1}{2})^5$ **C** ${}_{10}C_5 \times (\frac{1}{2})^{10}$ **D** $5 \times (\frac{1}{2})^5$ **E** $5 \times (\frac{1}{2})^5 \times (\frac{1}{2})^5$

Statement C is correct. The result of 5 heads implies 5 tails. These can be arranged in ${}_{10}C_5$ ways. ANSWER C

Multiple choice questions (Type B) Answer according to the table

A	B	C	D	E
1, 2, 3 correct	1, 3 only	2, 3 only	2 only	3 only

Example 8
$p(A) = 0.4$, $p(B) = 0.3$, $p(A \text{ or } B) = 0.6$

1 $p(A \cap B) = 0.7$ **2** $p(A \cap \bar{B}) = 0.3$ **3** $p(A \mid B) = \frac{1}{3}$

$$\underset{0.4}{p(A)} + \underset{0.3}{p(B)} = p(A \cap B) + \underset{0.6}{p(A \cup B)}$$

where $P(A \cap B)$ is probability of A and B and $p(A \cup B)$ is probability of A or B or both.

$\Rightarrow p(A \cap B) = 0.1$ so statement 1 is false. This result is similar to the result for the intersection and union of sets.

$p(A \cap \bar{B}) = p(A) - p(A \cap B) = 0.4 - 0.1 = 0.3$ Statement 2 is true

$$p(A \mid B) = \frac{p(A \cap B)}{p(B)} = \frac{0.1}{0.3} = \frac{1}{3}$$ Statement 3 is true

ANSWER C

Example 9
In throwing 2 dice, S is the event of a total score of 7 and E is the event 'only one of the numbers is even'.

1 $p(S) = \frac{7}{36}$ **2** $p(E) = \frac{1}{2}$ **3** S and E are independent

There are 36 possible results when 2 dice are thrown. Six of these, (1, 6), (6, 1), (2, 5), (5, 2), (3, 4), (4, 3), give a total of 7 so

$p(S) = \frac{6}{36} = \frac{1}{6}$ Statement 1 is wrong

Statement 2 is true, there being 18 pairs in which only 1 number is even.

$p(E \mid S) = \frac{6}{6} = 1$; $P(E \mid \bar{S}) = \frac{12}{30} = \frac{2}{5}$ so E and S are not independent.
$p(S \mid E) = \frac{6}{18} = \frac{1}{3}$; $p(S \mid \bar{E}) = 0$, which confirms this.
Statement 2 only is true. ANSWER D

> **Example 10**
> In drawing a card from a pack of 52 playing cards, the probability of drawing
>
> **1** a red card or an ace is $\frac{7}{13}$ **2** a red card or a black card is $\frac{1}{2}$ **3** a jack is $\frac{1}{13}$.

There are 26 red cards (including 2 red aces) and a further 2 black aces. The probability of a red card or an ace is $\frac{28}{52} = \frac{7}{13}$.
The probability of a red card is $\frac{1}{2}$ and the probability of a black card is $\frac{1}{2}$, but whichever card is drawn must be red or black, so the probability of a red or black card will be 1.

There are 4 jacks in the pack so the probability of a jack is $\frac{4}{52} = \frac{1}{13}$.
Statements 1 and 3 are true. ANSWER B

> **Example 11**
> The probability of success in a single trial is a. The probability of failure is b. In a series of n such independent trials the distribution of the number of successes has mean m and standard deviation s.
>
> **1** $a+b=1$ **2** $m = na$ **3** $s^2 = nab$

Each trial must result in either success or failure, so statement 1 is true. The probability generator for n trials is $G(t) = (b+at)^n$ where the coefficients of the powers of t in the expansion give the probabilities of the number of successes.

$$m = G'(1) = na(b+a)^{n-1} = na$$

$$\begin{aligned} s^2 = G'(1)+G''(1)-m^2 &= na+n(n-1)a^2(b+a)^{n-2}-m^2 \\ &= na+n(n-1)a^2-n^2a^2 \\ &= na+n^2a^2-na^2-n^2a^2 \\ &= na-na^2 = na(1-a) = nab \end{aligned}$$

Statements 1, 2 and 3 are all true. ANSWER A

> **Example 12**
> If 6 dice are thrown together
>
> **1** the expected number of sixes is 1 **2** the probability of one six is $(\frac{1}{6}) \times (\frac{5}{6})^6$ **3** the probability of no sixes is $(\frac{5}{6})^6$

The expected number of sixes is the mean number of sixes and 6 die thrown together can be regarded as 1 dice thrown 6 times. From Example 11, the mean number is $6 \times (\frac{1}{6}) = 1$.
Statement 1 is true

If one six has occurred so have 5 non-sixes giving a probability of $(\frac{1}{6}) \times (\frac{5}{6})^5$, but this can occur in any one of six ways (the six can occur on any one of the 6 dice) so the total probability is $6 \times (\frac{1}{6}) \times (\frac{5}{6})^5$. Statement 2 is wrong.
No sixes imply all non-sixes, i.e., $(\frac{5}{6})^6$.
Statement 3 is correct. ANSWER B

Example 13
Five letters are chosen from the word FAVOURITES. The probability that

1 they are all vowels is $\frac{1}{2}$ **2** they are all consonants is $\dfrac{1}{{}_{10}C_5}$

3 there are more vowels than consonants is $\frac{1}{2}$

The probability of all vowels or all consonants must be the same as there are 5 of each. There is only one way of choosing 5 vowels out of ${}_{10}C_5$ ways of choosing 5 letters so statement 1 is wrong and statement 2 is correct. In selecting 5 letters all the letters are different so each selection including 1 vowel contains 4 consonants, and 2 vowels gives 3 consonants, etc. Half the selections contain more vowels than consonants and half do not; statement 3 is correct. ANSWER C

General questions

Example 14
Two unbiased dice are thrown and the sum, S, is recorded on each throw. Find the mean and standard deviation of S.

The values for S range from 2 to 12 with the following probabilities

S	2	3	4	5	6	7	8	9	10	11	12
Probability	$\frac{1}{36}$	$\frac{2}{36}$	$\frac{3}{36}$	$\frac{4}{36}$	$\frac{5}{36}$	$\frac{6}{36}$	$\frac{5}{36}$	$\frac{4}{36}$	$\frac{3}{36}$	$\frac{2}{36}$	$\frac{1}{36}$

The mean m of S is $m = \Sigma S \times \text{Prob}(S) = 2 \times \frac{1}{36} + 3 \times \frac{2}{36} + 4 \times \frac{3}{36} + \ldots$

$$m = (2+6+12+20+30+42+40+36+30+22+12) \div 36$$
$$= 252 \div 36 = 7$$

This result is to be expected as the probability distribution is symmetrical.

The standard deviation is given by $\text{S.D.}^2 = \Sigma(S-m)^2 pr(S)$

$$\text{S.D.}^2 = (2-7)^2 \times \tfrac{1}{36} + (3-7)^2 \times \tfrac{2}{36} + (4-7)^2 \times \tfrac{3}{36} + \ldots$$
$$= 2\times 5^2 \times \tfrac{1}{36} + 2\times 4^2 \times \tfrac{2}{36} + 2\times 3^2 \times \tfrac{3}{36} + 2\times 2^2 \times \tfrac{4}{36}$$
$$+ 2\times 1^2 \times \tfrac{5}{36} + 0$$
$$= \tfrac{1}{18}[25+32+27+16+5] = \tfrac{105}{18} = 5\tfrac{5}{6} = 5.83$$
$$\text{S.D.} = \sqrt{5.83} \simeq 2.42$$

There are short cuts to calculations like this but in this case where the distribution is symmetrical (the mean being 7) it is just as quick to use the basic definitions of mean and S.D.

> **Example 15**
> On my journey to work 20% of the cars I see contain more than one person. On a particular morning, what is the probability that three of the first five cars I see contain more than one person?

The Binomial Probability Generating Function is $(\tfrac{4}{5}+\tfrac{1}{5}t)^5$.
For three cars the coefficient of t^3 is needed, i.e.

$$p(\text{3 cars with more than 1 person}) = 10(\tfrac{4}{5})^2 \times (\tfrac{1}{5})^3 = \frac{160}{5^5} = \frac{32}{5^4} \simeq 0.0512$$

The probability of seeing 3 or more cars with more than 1 person is

$$p(3)+p(4)+p(5) = 10(\tfrac{4}{5})^2\,(\tfrac{1}{5})^3 + 5(\tfrac{4}{5})\,(\tfrac{1}{5})^4 + 1(\tfrac{4}{5})^5 = \frac{160+20+1}{5^5}$$
$$= \frac{181}{5^5} \simeq 0.058$$

> **Example 16**
> $p(X) = \tfrac{1}{5} \quad p(Y\,|\,X) = \tfrac{1}{4} \quad p(Y\,|\,\bar{X}) = \tfrac{1}{3}$ find $p(X\,|\,Y)$

$$p(X \cap Y) = p(X) \times p(Y\,|\,X) = \tfrac{1}{5} \times \tfrac{1}{4} = \tfrac{1}{20}$$
$$p(\bar{X} \cap Y) = p(\bar{X}) \times p(Y\,|\,\bar{X}) = \tfrac{4}{5} \times \tfrac{1}{3} = \tfrac{4}{15}$$
$$p(Y) = p(X \cap Y) + p(\bar{X} \cap Y) = \tfrac{1}{20} + \tfrac{4}{15} = \tfrac{3+16}{60} = \tfrac{19}{60}$$
$$p(X\,|\,Y) = \frac{p(X \cap Y)}{p(Y)} = \frac{\tfrac{1}{20}}{\tfrac{19}{60}} = \tfrac{3}{19}$$

> **Example 17**
> Given $p(A) = \tfrac{2}{5}$, $p(B) = \tfrac{1}{4}$ and $p(A \cap B) = \tfrac{3}{20}$, find out whether A and B are statistically independent.

We need to find the conditional probabilities $p(A \mid B)$ and $p(A \mid \bar{B})$ or $p(B \mid A)$ and $p(B \mid \bar{A})$.

Method 1 Tree diagram. Using the notation in the diagram, (see Figure 26(a))

$$p(A) \times p(B \mid A) = p(A \cap B) \Rightarrow \tfrac{2}{5} \times b_1 = \tfrac{3}{20} \Rightarrow b_1 = p(B \mid A) = \tfrac{3}{8}$$
$$p(B) = p(A \cap B) + p(\bar{A} \cap B) \Rightarrow \tfrac{1}{4} = \tfrac{3}{20} + \tfrac{3}{5}b_2 \Rightarrow \tfrac{3}{5}b_2 = \tfrac{1}{10}$$
$$\Rightarrow b_2 = p(B \mid \bar{A}) = \tfrac{1}{6}$$

$p(B \mid A) \neq p(B \mid \bar{A}) \Rightarrow A$ and B are not independent. They are dependent.

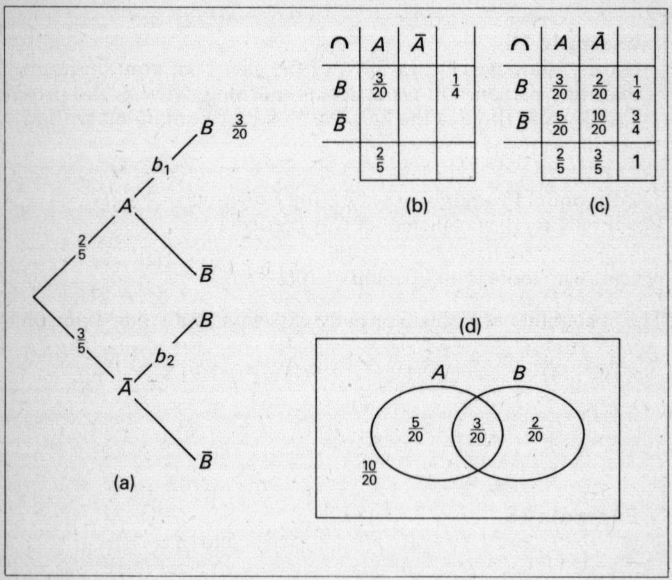

Figure 26

Method 2 Contingency table. In Figure 26(b) the information given is recorded. On the top line, since $p(B) = \tfrac{1}{4} = p(A \cap B) + p(\bar{A} \cap B)$, the missing value $p(B \cap \bar{A})$ is calculated as $\tfrac{1}{4} - \tfrac{3}{20} = \tfrac{2}{20}$. Similarly the rest of the table is calculated (see Figure 26(c)).

$$p(B \mid A) = \frac{\frac{3}{20}}{\frac{2}{5}} = \tfrac{3}{8} \text{ (1st column)} \quad p(B \mid \bar{A}) = \frac{\frac{2}{20}}{\frac{3}{5}} = \tfrac{1}{6} \text{ (2nd column)}$$

Method 3 Venn diagram.

Overlap $\equiv p(A \cap B) = \frac{3}{20} \Rightarrow$ LHS $= p(A \cap \bar{B}) = \frac{2}{5} - \frac{3}{20} = \frac{5}{20}$

and RHS $= p(B \cap \bar{A}) = \frac{1}{4} - \frac{3}{20} = \frac{2}{20} \Rightarrow$ Outside $= p(\bar{A} \cap \bar{B}) = \frac{10}{20}$ (see Figure 26(d)).

$p(B \mid A) = \dfrac{\frac{3}{20}}{\frac{8}{20}} = \frac{3}{8}$ and $p(B \mid \bar{A}) = \dfrac{\frac{2}{20}}{\frac{12}{20}} = \frac{1}{6} \Rightarrow A$ and B are dependent.

Example 18

I belong to a 100 Club, where 100 members pay in £1 per month. There are prizes of £24 and £12 each month except in June when there is one prize of £100 and in December when there is one prize of £200. What are my chances of breaking even (winning at least £12 during the year)?

The probability of winning £200 is 1 in 100, i.e., 0.01.
The probability of winning £100 is 1 in 100, i.e., 0.01.

In each of the other months the probability of winning £24 is 1 in 100 and the probability of winning £12 is 1 in 99. This happens 10 times during the year, so the probability of winning £12 or £24 during the year is $10(\frac{1}{100} + \frac{1}{99}) = \frac{1990}{9900}$.

Total probability of breaking even (at least I may win more than 1 prize)

$$= \frac{1990}{9900} + \frac{1}{100} + \frac{1}{100} = \frac{2188}{9900} = 0.221$$

I have almost a one in four chance of breaking even. How does the Club make money? With 100 members having a 1 in 4 chance of breaking even is there a profit?

The answer is that the probability of each member winning is not independent from each other. Not every member can win. In fact there are only 22 prizes and yet 100 members.

The 100 Club is superficially attractive to each member and yet it pays out £660 in prizes and makes £540 profit per year.

Example 19

If the probability of passing each driving test is $\frac{1}{3}$, find the mean number of tests each person takes.

The probability of taking 1 test is $\frac{1}{3}$.
The probability of taking 2 tests is $\frac{2}{3} \times \frac{1}{3}$ (fail the 1st, pass the 2nd).
The probability of taking 3 tests is $(\frac{2}{3})^2 \times \frac{1}{3}$ (fail 1st and 2nd, pass 3rd).

Probability generator $G(t) = \frac{1}{3}t + \frac{2}{3}\times\frac{1}{3}t^2 + (\frac{2}{3})^2\times\frac{1}{3}t^3 + \ldots = \dfrac{\frac{1}{3}t}{1-\frac{2}{3}t}$.

(Summing this as a converging infinite G.P.)

where the probability of taking n tests is the coefficient of t^n. The mean number of tests is given by

$$1\times\tfrac{1}{3} + 2\times\tfrac{2}{3}\times\tfrac{1}{3} + 3\times(\tfrac{2}{3})^2\times\tfrac{1}{3} + \ldots = G'(1)$$

$$G'(t) = \frac{d}{dt}\left(\frac{t}{3-2t}\right) = \frac{(3-2t)1 - t(-2)}{(3-2t)^2} = \frac{3}{(3-2t)^2} \Rightarrow G'(1) = 3$$

Mean number of tests taken is 3.

Example 20

At Esher College, 2 members of staff out of 50 have arrived by 8.30 a.m. and 15 by 8.40 a.m. Assuming the arrival times for members of staff to form a normal distribution, calculate the mean arrival time and find out how many staff are late for the first period which starts at 9.05 a.m.

If 8.30 a.m. corresponds to (mean $- x_1$)

$$\phi(+x_1) = 1 - \tfrac{2}{50} = 0.96$$

(This is the proportion of the population less than x_1 S.D.s to the right of the mean.)

$$x_1 = 1.751 \text{ S.D.s}$$

Similarly, if 8.40 a.m. corresponds to (mean $- x_2$) S.D.s

$$\phi(+x_2) = 1 - \tfrac{15}{50} = 0.7 \Rightarrow x_2 = 0.8418 \text{ S.D.s}$$

10 min. $\equiv$ 1.751 $-$ 0.8418 S.D.s $=$ 0.9092 S.D.s $\Rightarrow$ 1 S.D. $=$ 11 min.
0.8418 S.D.s $\equiv$ 0.8418 $\times$ 11 min. $=$ 9.259 min.
so mean $=$ 8 h. 49.259 min.

9.05 a.m. is 15.741 min. beyond mean, i.e., $\dfrac{15.741}{9.259}$ S.D.s $=$ 0.588 S.D.s

$\phi(0.588) = 0.7208$, i.e., 0.2792 of the population still to arrive.
0.2792 $\times$ 50 $=$ 13.96 members of staff. 14 late.
Is this a good probability model for the situation?

Chapter 13
Groups and Algebraic Structure

Multiple choice questions (Type A) Select the correct answer

> **Example 1**
> For the set of integers which of the following operations is not closed?
>
> **A** addition **B** subtraction **C** multiplication
> **D** division **E** taking the larger number

For the set of integers (whole numbers, both positive and negative, and zero) take two numbers and see whether the result of each operation produces an integer.

Answers A, B and C will always give another integer, e.g., $2+-1=1$; $2--3=5$; $2\times-3=-6$. Taking the larger number (E), will always give one of the numbers you started with, i.e., an integer.

However $2\div-3=-\frac{2}{3}$ which is a fraction and belongs to the set of rational numbers. Therefore, answer D is not closed.

ANSWER D

> **Example 2**
> Which of the following sets of integers do not form a group under addition?
>
> **A** integers reduced modulo 4 **B** modulo 5 **C** modulo 6
> **D** modulo 7 **E** they all form groups

Three of the combination tables are shown in Figure 27.
Each is closed, associative, has identity 0 and inverses for every element. The integers form a group under addition for all moduli.

ANSWER E

modulo 4

+	0	1	2	3
0	0	1	2	3
1	1	2	3	0
2	2	3	0	1
3	3	0	1	2

modulo 5

+	0	1	2	3	4
0	0	1	2	3	4
1	1	2	3	4	0
2	2	3	4	0	1
3	3	4	0	1	2
4	4	0	1	2	3

modulo 6

+	0	1	2	3	4	5
0	0	1	2	3	4	5
1	1	2	3	4	5	0
2	2	3	4	5	0	1
3	3	4	5	0	1	2
4	4	5	0	1	2	3
5	5	0	1	2	3	4

Figure 27

Example 3
For which of the following values do the integers, reduced modulo p, **not** form a group under multiplication modulo p.

A $p = 3$ **B** $p = 5$ **C** $p = 7$ **D** $p = 9$ **E** $p = 11$

A group is formed for each value as long as p is prime, so answer D is the only value which does not form a group. The reason is that in D we have divisors of zero. This means that there are elements which multiply together to form zero, e.g., $3 \times 3 = 0$, $3 \times 6 = 0$. The set is not closed, zero being excluded. Even if zero were included, 0 would not have an inverse. The identity is 1 and there is no element which combines with 0 to produce 1 so 0 has no inverse. This does not occur when p is prime since in that case p can have no factors.

ANSWER D

Example 4
Which of the following systems is not distributive?

A multiplication over addition for real numbers
B multiplication over subtraction for real numbers
C addition over multiplication for real numbers
D union over intersection for sets
E intersection over union for sets

The operation $*$ is distributive over @ if $a * (b \,@\, c) = (a * b) \,@\, (a * c)$

Answers A and B are true, e.g., $5 \times (4 \pm 3) = (5 \times 4) \pm (5 \times 3)$.
$5 + (3 \times 4) = 17$ but $(5+3) \times (5+4) = 72$, so addition is not distributive over multiplication.
ANSWER C

Answers D and E are both true. For example, $X \cup (Y \cap Z) = (X \cup Y) \cap (X \cup Z)$. This is best demonstrated by shading the respective sets on a Venn diagram.

Example 5
Which of the following equations has no solution if $x \in Z_5$, the set of integers modulo 5?

A $2x = 3$ **B** $2x+3=0$ **C** $2(x+3)=0$ **D** $x^2+1=0$
E $x^2+2=0$

A $2x = 3 \Rightarrow x = 4$ since $2 \times 4 = 8 - 3$ (modulo 5)
B $2x+3 = 0 \Rightarrow 2x = 2 \Rightarrow x = 1$
C $2(x+3) = 0 \Rightarrow x+3 = 0 \Rightarrow x = 2$
D $x^2+1 = 0 \Rightarrow x^2 = 4 \Rightarrow x = 2$ or 3 since $3^2 = 4$
E $x^2+2 = 0 \Rightarrow x^2 = 3 \Rightarrow$ no solution

$1^2 = 1, 2^2 = 4, 3^2 = 4, 4^2 = 1$, so $x = 3$ has no solution.
ANSWER E

In D, $x^2+1 = x^2+5x+6 = (x+2)(x+3) = 0$
$x+2 = 0 \Rightarrow x = 3$
$x+3 = 0 \Rightarrow x = 2$

Multiple choice questions (Type B) Answer according to the table

A	**B**	**C**	**D**	**E**
1, 2, 3 correct	1, 3 only	2, 3 only	2 only	3 only

Example 6
The set of matrices of the form $\begin{pmatrix} a & -b \\ b & a \end{pmatrix}$ under the operation of matrix multiplication

1 is closed **2** is commutative **3** is associative

To examine the property of closure consider two matrices of this type, i.e.,

$$\begin{matrix} A & \times & C & = & AC \end{matrix}$$
$$\begin{pmatrix} a & -b \\ b & a \end{pmatrix}\begin{pmatrix} c & -d \\ d & c \end{pmatrix} = \begin{pmatrix} ac-bd & -ad-bc \\ bc+ad & -bd+ac \end{pmatrix}$$

Since in the product AC the components on the leading diagonal are equal and those on the reverse diagonal are equal and opposite in sign, the product matrix has the same form as A and C. Make sure that the components of each matrix a, b, c and d belong to the same set of numbers. For instance the question may have specified that they be Z(integers) or Q(rationals) or R(real numbers) in which case the elements of the product AC must belong to these sets also. In fact, in each of these cases this is true. The set is closed.

To investigate commutativity evaluate CA

$$\begin{matrix} C & \times & A & = & CA \end{matrix}$$

$$\begin{pmatrix} c & -d \\ d & c \end{pmatrix}\begin{pmatrix} a & -b \\ b & a \end{pmatrix} = \begin{pmatrix} ac-bd & -bc-da \\ ad+bc & -db+ac \end{pmatrix} = AC$$

$AC = CA \Rightarrow$ commutativity (which is not true in general for matrix multiplication) so the operation is commutative.

Matrix multiplication is associative for all compatible matrices. In this case for $E = \begin{pmatrix} e & -f \\ f & e \end{pmatrix}$ evaluate $(AC)E$ and $A(CE)$, the brackets indicating which pair to work out first.
Statements 1, 2 and 3 are all correct. ANSWER A

This question can also be answered with a little knowledge of transformations. If the determinant of these matrices were equal to 1, each matrix would represent a rotation in two dimensions about the origin. In that case the properties of closure, commutativity and associativity are easily verified. As it is, matrix A represents a rotation (of angle $\arctan b/a$) combined with an enlargement of area scale factor $\Delta = a^2 + b^2$. This combination of transformations is called a spiral similarity. The order of performing the rotation and enlargement does not affect the result. Knowing that rotations and enlargements are closed, associative and commutative could easily give the results above.

Example 7
Which of the following combination tables represents a group structure?

1

*	a	b	c	d
a	a	b	c	d
b	d	a	b	c
c	c	d	a	b
d	b	c	d	a

2

*	a	b	c
a	a	b	c
b	b	c	a
c	c	a	b

3

*	a	b	c
a	a	b	c
b	c	a	b
c	b	c	a

Combination table 2 is the cyclic group of order 3. It is isomorphic (same structure) to the integers modulo 3 under addition, and also to rotations of 120°, 240° and 360° and, as such, forms a group. When the operation is unspecified it is easier to prove a group by establishing an isomorphic structure, and quoting the result for the isomorphic structure.

Combination tables 1 and 3 have no proper identity. For both tables, a is a left-side identity, i.e., $a * b = b$, $a * c = c$, $a * d = d$ in table 1, but no right-side identity exists. In both these cases the operation is not associative.

For 1, $b(cd) = bb = a$ but $(bc)d = bd = c$
For 3, $b(ac) = bc = b$ but $(ba)c = cc = a$ ANSWER D

> **Example 8**
> For $a,b \in R$ (real numbers) $a * b = \frac{1}{2}(a+b)$. The operation:
>
> **1** is associative **2** is commutative **3** has an identity

$b * a = \frac{1}{2}(b+a) = \frac{1}{2}(a+b)$, so the operation is commutative and statement 2 is true.

$$\left.\begin{aligned}(a*b)*c &= \tfrac{1}{2}[\tfrac{1}{2}(a+b)+c] = \tfrac{1}{4}a+\tfrac{1}{4}b+\tfrac{1}{2}c \\ a*(b*c) &= \tfrac{1}{2}[a+\tfrac{1}{2}(b+c)] = \tfrac{1}{2}a+\tfrac{1}{4}b+\tfrac{1}{4}c\end{aligned}\right\}\text{ These are only the same when } a = c.$$

Therefore statement 1 is true.
If b is the identity element $\frac{1}{2}(a+b) = a \Rightarrow \frac{1}{2}b = \frac{1}{2}a$ or $a = b$.
In general the only element which combines with a to produce a is a itself, but an identity has to be the same for every element of the set. Therefore, there is no identity element and so statement 3 is false. ANSWER D

> **Example 9**
> The elements A and B generate a group where $A^4 = I$ (identity), $A^2 = B^2$ and $AB = BA^3$.
>
> **1** $AB = BA$ **2** $A^2B = AB^2$ **3** $A^3B = BA$

$AB = BA \Rightarrow BA = BA^3 \Rightarrow A^2 = I \Rightarrow$ Statement 1 is not true.

$$\begin{aligned}AB = BA^3 \Rightarrow AB^2 &= BA^3B = BA^2AB = BA^2BA^3 \\ &= BAABA^3 = BABA^3A^3 \\ &= BABA^2 = BBA^3A^2 = B^2A\end{aligned}$$

Statement 2 is true

This is achieved by changing each AB into BA^3

$A^3B = A^2AB = B^2BA^3 = BB^2AA^2 = BAB^2A^2 = BAA^2A^2 = BA$
Statement 3 is true and statements 2 and 3 are true. ANSWER C

General questions

> **Example 10**
> Show that the set of non-zero real numbers forms a group under the operation $a * b$ where $a * b = 2ab$.

$a, b \in R$(real numbers) $\Rightarrow 2ab \in R$ so the set is closed.

$$(a * b) * c = (2ab) * c = 4abc \qquad a * (b * c) = a * (2bc) = 4abc$$

The operation is associative.
If b is the identity element, $a * b = a \Rightarrow 2ab = a \Rightarrow b = \frac{1}{2}$.
$b * a = \frac{1}{2} * a = 2 \cdot \frac{1}{2}a = a$, so $a * \frac{1}{2} = \frac{1}{2} * a = a$ and $\frac{1}{2}$ is the identity.
If b is the inverse of a, $a * b = \frac{1}{2} \Rightarrow 2ab = \frac{1}{2} \Rightarrow b = \frac{1}{4}a$. Each element a has an inverse $\frac{1}{4}a$, and the set forms a group.

> **Example 11**
>
> Find the group generated by the matrix $M = \begin{pmatrix} 0 & 1 & 0 \\ 0 & 0 & 1 \\ 1 & 0 & 0 \end{pmatrix}$

$$M^2 = \begin{pmatrix} 0 & 1 & 0 \\ 0 & 0 & 1 \\ 1 & 0 & 0 \end{pmatrix}\begin{pmatrix} 0 & 1 & 0 \\ 0 & 0 & 1 \\ 1 & 0 & 0 \end{pmatrix} = \begin{pmatrix} 0 & 0 & 1 \\ 1 & 0 & 0 \\ 0 & 1 & 0 \end{pmatrix}$$

$$M^3 = \begin{pmatrix} 0 & 0 & 1 \\ 1 & 0 & 0 \\ 0 & 1 & 0 \end{pmatrix}\begin{pmatrix} 0 & 1 & 0 \\ 0 & 0 & 1 \\ 1 & 0 & 0 \end{pmatrix} = \begin{pmatrix} 1 & 0 & 0 \\ 0 & 1 & 0 \\ 0 & 0 & 1 \end{pmatrix} = I$$

The matrix M generates the cyclic group of order 3.
This method may be tedious if the group has order 6 or more.
It may be easier to consider the transformation that M represents.

M takes $I(1,0,0)$ to $K(0,0,1)$; $J(0,1,0)$ to $I(1,0,0)$ and $K(0,0,1)$ to $J(0,1,0)$.

This rotates the triangle IJK to JKI, i.e., a rotation of 120°. M^2 will represent a rotation of 240° and $M^3 = I$ (rotation of 360°). The rotations form a group of order 3 and so will the matrices.

> **Example 12**
> Find the group with the smallest order which contains the complex number i where elements are complex numbers under the operation of multiplication.

The smallest group will be that generated by i; in other words all the elements will be powers of i.

$$i^2 = -1;\ i^3 = -i,\ i^4 = 1$$

So the group consists of $(1, i, -1, -i)$. The set is closed and associative; 1 is the identity; 1 and -1 are self-inverse and i and $-i$ form an inverse pair.
As an extra question, show that this group is cyclic.

Example 13
Find the order of each element of the group of integers under multiplication modulo 7 and use this to show that the group is cyclic.

The order of an element is the value of the lowest power which gives the identity 1.

$2^2 = 4$, $2^3 = 1$, so the order of 2 is 3.
$3^2 = 2$, $3^3 = 6$, $3^4 = 4$, $3^5 = 5$, $3^6 = 1$ and the order of 3 is 6.
$4^2 = 2$, $4^3 = 1$; the order of 4 is 3
$5^2 = 4$, $5^3 = 6$, $5^4 = 2$, $5^5 = 3$, $5^6 = 1$; the order of 5 is 6
$6^2 = 1$; the order of 6 is 2

Each element generates a subgroup. 3 and 5 generate the complete group so if the elements were written in the order of the powers of either 3 or 5, the group would be seen to be cyclic (see Figure 28).

MULTIPLICATION MODULO 7

×	1	2	3	4	5	6
1	1	2	3	4	5	6
2	2	4	6	1	3	5
3	3	6	2	5	1	4
4	4	1	5	2	6	3
5	5	3	1	6	4	2
6	6	5	4	3	2	1

×	1	3	3^2	3^3	3^4	3^5
1	1	3	2	6	4	5
3	3	2	6	4	5	1
$3^2 = 2$	2	6	4	5	1	3
$3^3 = 6$	6	4	5	1	3	2
$3^4 = 4$	4	5	1	3	2	6
$3^5 = 5$	5	1	3	2	6	4

Figure 28

Example 14
Show that the permutations of three letters (A,B,C) form a group.

There are $3\times2\times1=6$ ways of arranging ABC so there are 6 permutations of the three letters. When performed on the order ABC we get

$$ABC \quad ACB \quad BAC \quad BCA \quad CAB \quad CBA$$

These correspond to transforming the equilateral triangle as in Example 17 (Figure 30). The permutations are in one–one correspondence with the transformations and so they form a group. The permutations corresponding to the rotations BCA and CAB together with the identity ABC are **even** permutations, and ACB, BAC and CBA, corresponding to the opposite isometries, (reflexions) are odd permutations.

If you can show that a system is isomorphic to another system then the properties of each carry over.

Example 15
Examine the binary operation $a*b=a+b+1$ for group properties.

Closure depends on which set of numbers a and b are taken from. For N, Z, R, Q and C the set is closed but for the set of even numbers it is not closed.

$$\left.\begin{aligned}(a*b)*c&=(a+b+1)*c=a+b+1+c+1\\ a*(b*c)&=a*(b+c+1)=a+b+c+1+1\end{aligned}\right\}\ \text{The operation is associative.}$$

If b is the identity, $a*b=a\Rightarrow a+b+1=a\Rightarrow b=-1$
If b is the inverse of a, $a*b=-1\Rightarrow a+b+1=-1\Rightarrow b=-2-a$
The inverse of a is $-2-a$; $(-2-a)*a=-2-a+a+1=-1$ also.
The sets Z, Q, R and C form groups for the operation.
N does not, as there is no identity element, nor are there any inverses.

Example 16
Show that each of the sets $A=(1,3,7,9)$ and $B=(2,4,6,8)$ form groups under multiplication modulo 10. Why is 5 excluded from both sets? Show that A and B are isomorphic.

The combination tables are shown in Figure 29(a) and (b).

MULTIPLICATION MODULO 10

×	1	3	9	7
1	1	3	9	7
3	3	9	7	1
9	9	7	1	3
7	7	1	3	9

(a)

×	2	4	6	8
2	4	8	2	6
4	8	6	4	2
6	2	4	6	8
8	6	2	8	4

(b)

×	6	2	4	8
6	6	2	4	8
2	2	4	8	6
4	4	8	6	2
8	8	6	2	4

(c)

Figure 29

The order 1,3,9,7 emphasizes that A is a cyclic group in which the order 1397 appears along each row but starting each time with the heading element. From each table both sets are closed.

The associative property is often difficult to justify. Some students just prove it for a particular combination of elements like 3, 7 and 9 in A by showing that $3 \times (9 \times 7) = (3 \times 9) \times 7$, but this is not enough. It must be proved to be true for all combinations of three elements in the set, i.e., for $4 \times 4 \times 4 = 4^3 = 64$ triples in all (although some of these triples are trivial).

A general proof is needed or in this case an argument that can be extended to apply to all triples in general.

All members of the set containing 3 can be expressed as $3 + 10n$

$$\begin{aligned} \text{So } (3 \times 9) \times 7 &= (3+10n) \times (9+10p) \times (7+10q) \quad n,p,q \text{ are integers} \\ &= (27+90n+30p+100pn) \times (7+10q) \quad (1) \\ &= (3 \times 9) \times 7 + \text{multiples of } 10 \end{aligned}$$

Similarly $3 \times (9 \times 7) = 3 \times (9 \times 7) +$ multiples of $10 = 3 \times 9 \times 7$ (modulo 10). Since $3 \times (9 \times 7) = (3 \times 9) \times 7$ for integers the same result holds for integers reduced modulo 10, because all the terms neglected from (1) are multiples of 10. This argument can be applied to every single combination of three elements from set A and set B so in both tables the operation is associative.
The identity in A is 1 and in B is 6.
In A, 1 and 9 are self-inverse and 3 and 7 form an inverse pair.
In B, 6 and 4 are self-inverse and 2 and 8 form an inverse pair.

Both sets A and B form groups, the four conditions being satisfied. If 5 were a member of set B its combination with each element would produce 0. The set would not be closed. Even if 0 were included it would not have an inverse since there is no number which combines with 0 to give the identity 6.

If 5 were included in A, its combination with any of 1, 3, 7 or 9 would produce 5, and 5 would have no inverse. In both cases 5 behaves like a zero element.

In Figure 29(a) the table has been constructed in the order 1,3,9,7 to emphasize the cyclic property. In Figure 29(c) the order 6,2,4,8 produces the cyclic group of order 4 so the 2 groups have the same structure, i.e., they are isomorphic.

Example 17

Derive the group generated by the functions $f(x) = \frac{1}{x}$ and $g(x) = 1 - x$ under the combination of functions law where fg means g followed by f assuming that the combination of functions is associative.

First notice that $f^2 = ff(x) = x$ and call this the identity $I(x) = I$. Complete the combination table for the three elements I, f, g (see Figure 30(a)). Two new elements are generated, fg and gf, which are not equal.

$$fg(x) = \frac{1}{1-x} \quad \text{and} \quad gf(x) = 1 - \frac{1}{x} = \frac{x-1}{x}$$

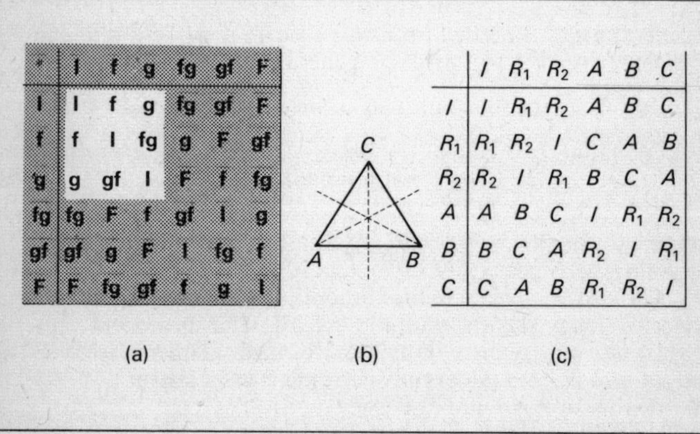

*	I	f	g	fg	gf	F
I	I	f	g	fg	gf	F
f	f	I	fg	g	F	gf
g	g	gf	I	F	f	fg
fg	fg	F	f	gf	I	g
gf	gf	g	F	I	fg	f
F	F	fg	gf	f	g	I

(a)

(b)

	I	R_1	R_2	A	B	C
I	I	R_1	R_2	A	B	C
R_1	R_1	R_2	I	C	A	B
R_2	R_2	I	R_1	B	C	A
A	A	B	C	I	R_1	R_2
B	B	C	A	R_2	I	R_1
C	C	A	B	R_1	R_2	I

(c)

Figure 30

These can be added to the table. Notice that $(gf)f = gf^2 = g$ since $f^2 = I$ and $f(fg) = f^2g = g$ since $g^2 = 1-(1-x) = x = I$.

$$g(gf) = g^2f = f \text{ and } (fg)g = fg^2 = f.$$

These results are easily deduced knowing the associative law holds.

$fgf(x) = \dfrac{1}{\dfrac{x-1}{x}} = \dfrac{x}{x-1}$: a new function;

$$gfg = 1 - \frac{1}{1-x} = \frac{1-x-1}{1-x} = \frac{-x}{1-x} = \frac{x}{x-1}$$

So $fgf = gfg = F$ and the rest of the group table can be derived algebraically.

For instance $(gf)(gf) = g(fgf) = g(gfg) = g^2fg = fg$
and $(fg)(fg) = f(gfg) = f(fgf) = f^2gf = gf$

The last row and column for F is now evaluated.

$fF = ffgf = gf$; $gF = ggfg = fg$; $Ffg = fgffg = fgg = f$
$fgF = fggfg = g$; $gfF = gffgf = f$; $Fgf = gfggf = g$
$FF = fgffgf = I$ or $gfggfg = I$

The last element F completes the set which forms a group of order 6. The group is not commutative as $fg \neq gf$.

Example 18
Form the group table for the symmetry operations of the square.

I is the identity, Q a quarter-turn (rotation 90°), H a half-turn and T a three quarters-turn all about the centre of the square. These form a commutative cyclic subgroup so the top left-hand corner is easily done. X and Y are reflexions in the horizontal and vertical lines of symmetry and M and N are reflexions in the diagonal lines of symmetry (see Figure 31).

*	I	Q	H	T	X	Y	M	N
I	I	Q	H	T	X	Y	M	N
Q	Q	H	T	I	M	N	Y	X
H	H	T	I	Q	Y	X	N	M
T	T	I	Q	H	N	M	X	Y
X	X	N	Y	M	I	H	T	Q
Y	Y	M	X	N	H	I	Q	T
M	M	X	N	Y	Q	T	I	H
N	N	Y	M	X	T	Q	H	I

The product of a direct and an opposite isometry is opposite so the top right- and bottom left-hand corners will consist of X, Y, M and N while the bottom right will consist of I, Q, H and T.

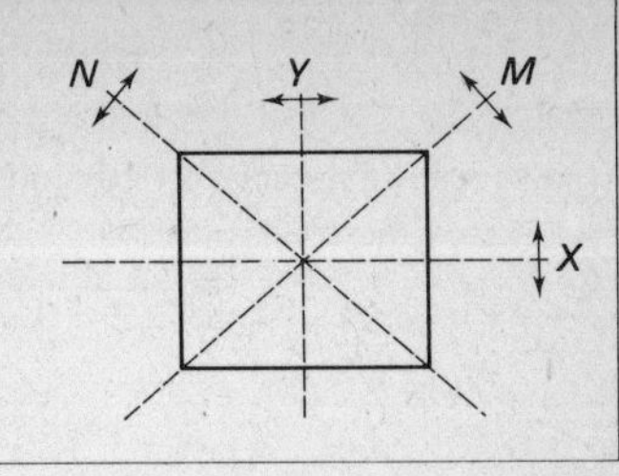

Figure 31

Since two reflexions give a rotation of double the angle between the mirrors $XY = YX = H$; also $HX = XH = Y$ and $HY = YH = X$ to give a subgroup (I,H,X,Y). This works for (I,H,M,N) since M and N are reflexions in mirrors at right angles to each other. MX means X first then M; two reflexions in mirrors at 45° = rotation of 90° = Q. XM will be the inverse of MX (since $XMMX = I$) so $MX = T$ and we now have the bottom right-hand corner knowing that there is one of each element in each row and column.

QX is X first then Q which gives $M \cdot XQ = N$
$QM = Y$ and $MQ = X$ and hence the complete table is obtained.

Chapter 14
Statics of a Particle

In this and succeeding chapters the following symbols will be used.
↑ to indicate resolving forces or considering motion vertically.
→ to indicate resolving forces or considering motion horizontally.
$\widehat{A}$ taking moments about the point A.

Multiple choice questions (Type A) Select the correct answer

> **Example 1**
> Forces $\mathbf{F}_1$ (1 Newton North), $\mathbf{F}_2$ (2 N South), $\mathbf{F}_3$ (3 N East) and $\mathbf{F}_4$ (4 N West) act on a particle. The magnitude of their resultant is
>
> **A** 10 **B** 2 **C** $\sqrt{2}$ **D** $5\sqrt{2}$ **E** $\sqrt{1^2+2^2+3^2+4^2} = \sqrt{30}$

Forces are added like vectors, their direction being important. The easiest way in this case is to express each force in terms of the unit vectors **i** and **j**. Thus

$\mathbf{F}_1 = \mathbf{j},\ \mathbf{F}_2 = -2\mathbf{j},\ \mathbf{F}_3 = 3\mathbf{i},\ \mathbf{F}_4 = -4\mathbf{i}$
$\mathbf{F}_1+\mathbf{F}_2+\mathbf{F}_3+\mathbf{F}_4 = -\mathbf{i}-\mathbf{j}$, which has magnitude $\sqrt{2}$ in direction SW.

ANSWER C

> **Example 2**
> A particle is in equilibrium with the three forces 3 N, 4 N and 5 N acting on it. The angle between the 3 N and 4 N forces is
>
> **A** 45° **B** 60° **C** 70° **D** 80° **E** 90°

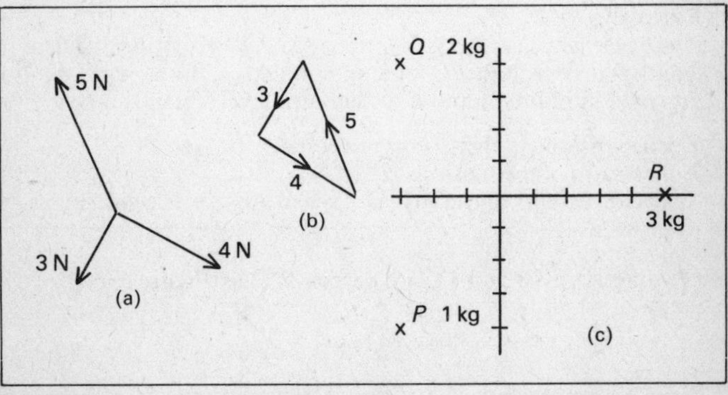

Figure 32

If the forces are in equilibrium (Figure 32(a)) they can be represented by the sides of a triangle (Figure 32(b)) which form a right-angled triangle (Pythagoras). ANSWER E

Example 3
Three particles P (of mass 1 kg), Q (2 kg) and R (3 kg) are placed at the points $(-3, -4)$, $(-3, 4)$ and $(5, 0)$ respectively. Their combined centre of mass is at the point

A $(0,0)$ **B** $(-1,0)$ **C** $(-\frac{1}{3},0)$ **D** $(1,\frac{2}{3})$ **E** $(6,4)$

The centre of mass is $\frac{1}{6}[1\times(-3,-4)+2\times(-3,4)+3\times(5,0)]$

$$=\tfrac{1}{6}[(-3-6+15, -4+8+0)] = \tfrac{1}{6}(6,4) = (1,\tfrac{2}{3})$$

ANSWER D

C would have been the correct answer if all the masses had been equal.

The centre of mass of 1 kg and 2 kg divides PQ in the ratio 2:1 so these are equivalent to a mass of 3 kg at $(-3, 1\frac{1}{3})$. The centre of mass of this together with 3 kg at R is half-way between $(-3, 1\frac{1}{3})$ and $(5, 0)$, i.e., at $(1, \frac{2}{3})$. See Figure 32(c).

Multiple choice questions (Type B) Answer according to the table

A	B	C	D	E
1, 2, 3 correct	1, 3 only	2, 3 only	2 only	3 only

Example 4
A particle of mass 10 kg is suspended by two strings: AB of length 3 metres, and BC, of length 4 metres, the strings being attached at points A and C at the same level 5 metres apart.

1 The tensions in the strings are equal.
2 The ratio of the tensions is 3:4.
3 The directions of the strings AB and AC are at right angles.

By Pythagoras, AB and BC are perpendicular (Figure 33(a)).

$$T_1 \cos x = T_2 \cos y \Rightarrow T_1 \times \tfrac{3}{5} = T_2 \times \tfrac{4}{5} \Rightarrow \frac{T_1}{T_2} = \tfrac{4}{3}$$

T_1 and T_2 are not equal but they are in the ratio $T_2 : T_1 = 3:4$. Statement 2 only is true. ANSWER D

Since AB and BC are perpendicular, the easiest way to find T_1 and T_2 is to resolve in the directions of AB and BC.

In direction AB $\quad T_1 = 10g \cos y = 10g \times \frac{4}{5} = 8g\,\text{N}$
In direction BC $\quad T_2 = 10g \cos x = 10g \times \frac{3}{5} = 6g\,\text{N}$

Alternatively, using Lami's theorem (three concurrent forces)

$$\frac{T_1}{\sin(180° - x)} = \frac{T_2}{\sin(180° - y)} = \frac{10g}{\sin 90°} \Rightarrow \begin{array}{l} T_1 = 10g \sin x = 8g \\ T_2 = 10g \sin y = 6g \end{array}$$

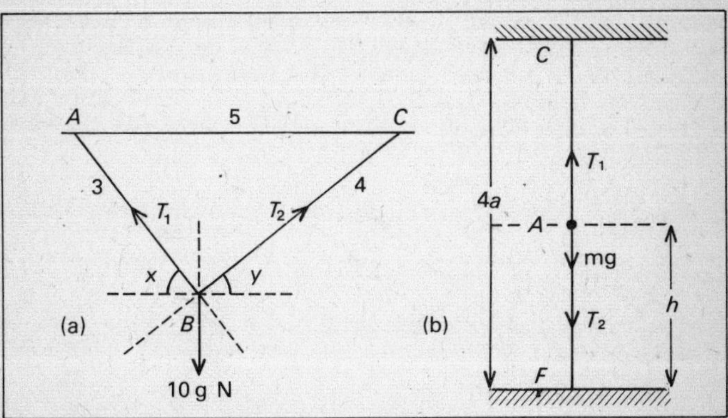

Figure 33

Example 5
Two elastic strings CA (modulus $2mg$ and natural length a) and AF (modulus mg and length a) are fixed together at A. The first string is fixed to the ceiling at C and the second string to the floor at F, where $CF = 4a$. A mass m is attached at A. Find the position of the mass if it is allowed to hang in equilibrium with the strings taut.

Let the equilibrium position of A be at a height h above F.

The extension of CA is $4a - h - a = 3a - h \Rightarrow T_1 = \dfrac{2mg}{a}(3a - h)$

The extension of AF is $h - a \Rightarrow T_2 = \dfrac{mg}{a}(h - a)$

$$T_1 = T_2 + mg \Rightarrow \frac{2mg}{a}(3a-h) = \frac{mg}{a}(h-a) + mg \Rightarrow 6a - 2h = h - a + a$$

$\Rightarrow 6a = 3h \Rightarrow h = 2a$ and A is half-way between C and F.

Example 6
Three forces are represented in magnitude and direction by the sides of a triangle (Figure 34(a)).

1 They are in equilibrium.
2 Their resultant is zero.
3 They form a couple.

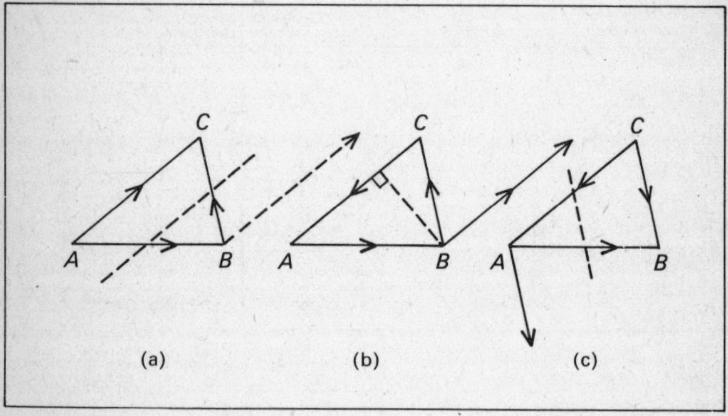

Figure 34

This question is deliberately worded too vaguely, to make several points. Firstly, the direction of the forces must be specified. Three different possibilities are given in Figure 34. Secondly, when forces are specified we must know whether they are acting on a particle or on a rigid body (in this case a triangular lamina) when their lines of action are important.

For a particle, if the three forces are in consecutive order round the triangle (as in Figure 34(b)) they balance and the particle is in equilibrium. In Figure 34(a), the resultant is 2**AC** and in Figure 34(c) it is 2**CB**.
For a rigid body (lamina) none are in equilibrium.

In Figure 34(a), the resultant of **AB** and **BC** is a force equal to **AC** acting through B. The resultant of this and **AC** will be a force equal to 2**AC** acting through the mid-points of AB and BC. Similarly, in Figure 34(c), the resultant is a force of 2**CB** acting through the mid-points of AC and AB.

In Figure 34(b), **AB** + **BC** give a force equal to **AC** acting through B so we have two equal and opposite forces which form a couple (whose magnitude is $|\mathbf{AC}| \times h$).

In this case statement 3 only is correct.

Where the forces act on a particle, statements 1 and 2 only are correct (see Figure 34(b)).

General questions

> **Example 7**
> Find the acceleration of masses 3 kg and 5 kg (which hang over a smooth pulley) connected by a light inextensible string. Find also the tension in the string.

A smooth pulley is one in which there is no friction at the bearings. In this case the tension in the string is the same throughout its length. The masses will not remain stationary. With the notation in Figure 35 let the 3 kg mass accelerate upwards with acceleration a and so the 5 kg mass will accelerate downwards with acceleration a.

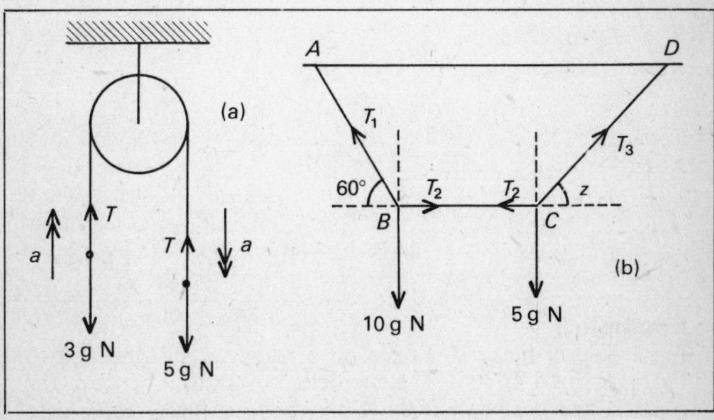

Figure 35

For the 3 kg mass, using $F = ma$ $\quad T - 3g = 3a \quad (1)$
For the 5 kg mass, using $F = ma$ $\quad 5g - T = 5a \quad (2)$

Adding (1) and (2) gives $2g = 8a \Rightarrow a = \frac{1}{4}g$
From (1) $\quad T = 3g + 3a = 3g + \frac{3}{4}g = 3\frac{3}{4}g$

> **Example 8**
> A string $ABCD$ is suspended from two points A and D at the same level with masses of 10 kg attached at B and 5 kg at C. The section BC is horizontal. Find the tensions in AB, BC and CD and the angle CD makes with the horizontal if angle BAD is 60° ($g = 9.8\,\text{m s}^{-2}$).

This is a straightforward question involving resolving forces at each point and solving the resulting equations. Consider each mass separately and identify the forces acting. Then resolve in appropriate directions, in this case horizontally and vertically.

With the notation in Figure 35(b)

$$\text{At } B \uparrow \; T_1 \sin 60° = 10g \Rightarrow T_1 = 10g \div \frac{\sqrt{3}}{2} = \frac{20g}{\sqrt{3}} \simeq 113\,\text{N}$$

$$\leftarrow T_1 \cos 60° = T_2 \;\Rightarrow T_2 = 113 \times \tfrac{1}{2} \quad = 56.5\,\text{N}$$

$$\text{At } C \; \left.\begin{array}{l} \uparrow T_3 \sin z = 5g \\ \leftarrow T_3 \cos z = 56.5 \end{array}\right\} \quad \tan z = \frac{5g}{56.5} = 0.8673 \Rightarrow z = 40.9°$$

$$T_3 = \frac{56.5}{\cos z} = \frac{56.5}{0.7555} = 74.8\,\text{N}$$

> **Example 9**
> A particle of mass 10 g rests on a rough plane (inclined to the horizontal at 30°). The coefficient of friction between the particle and the plane is 0.6. Find the least force required to move the particle up the plane and the angle at which it acts.

Friction opposes motion so F acts down the plane. If the particle is moving or is about to move then F is equal to the limiting friction $F = \mu R$. With the notation in Figure 36, let P be the required force (to move the particle up the plane) acting at an angle θ above the angle of the plane.

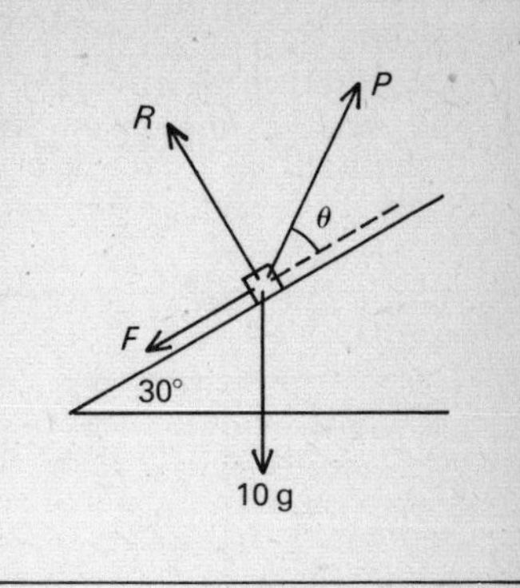

Figure 36

$\nearrow$ $P\cos\theta > F + 10g\sin 30°$

$= \mu R + 10g \times \frac{1}{2}$

$\nwarrow$ $P\sin\theta + R = 10g\cos 30°$

$\Rightarrow R = 10g\frac{\sqrt{3}}{2} - P\sin\theta$

$\Rightarrow P\cos\theta > \mu R + 5g$

$= 5\sqrt{3}\mu g - \mu P\sin\theta + 5g$

$P\cos\theta + \mu P\sin\theta > 5\sqrt{3}\mu g + 5g = 3\sqrt{3}g + 5g = 10.2g = 99.9\text{ N}$

Writing $\mu = \tan k = 0.6 \Rightarrow P(\cos\theta + \tan k\sin\theta) > 99.9\text{ N}$

$\Rightarrow P(\cos\theta\cos k + \sin\theta\sin k) = 99.9\cos k \Rightarrow P\cos(\theta - k) = 99.9\cos k$

$\Rightarrow P > \dfrac{99.9\cos k}{\cos(\theta - k)} = \dfrac{99.9\cos 30.96°}{\cos(\theta - k)}$

P is least when $\cos(\theta - k)$ is greatest, i.e., when $\theta = k = 30.96°$. In that case, $\cos(\theta - k) = 1$ and $P = 99.9\cos 30.96° = 85.7\text{ N}$.

Example 10
A book is placed on a desk lid, the coefficient of friction between the two being $\frac{1}{3}$. The desk lid is gradually raised until the book begins to slide. When this happens what is the angle the desk lid makes with the horizontal?

The only forces acting are the weight W, friction F and the normal reaction R.

On the point of moving $\nearrow$ $F = \mu R = W\sin\theta$ (θ is the angle of the lid)

$\nwarrow$ $R = W\cos\theta$

$\dfrac{F}{R} = \mu = \dfrac{W\sin\theta}{W\cos\theta} \Rightarrow \tan\theta = \frac{1}{3} \Rightarrow \theta = \tan^{-1}\frac{1}{3} = 18.4°$

Example 11
A particle C of mass 10 kg is suspended by two strings, AC and AB, each of length 1 m, from two points A and B on the same level where $AB = 1$ m. Find the tensions in the strings and the level of C below AB if (a) the strings are inelastic, (b) both strings are elastic of modulus $5g$ N and length 1 m.

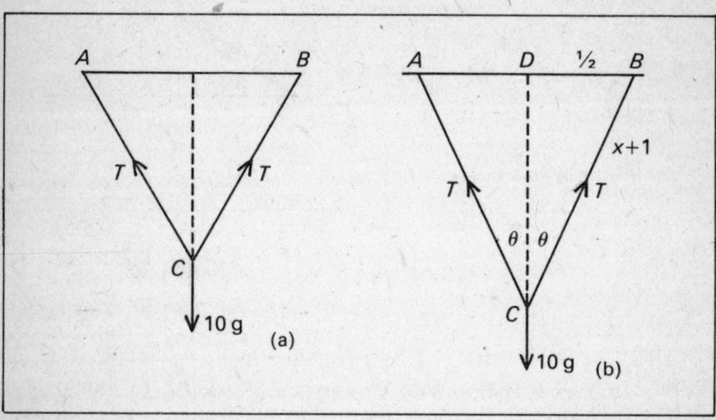

Figure 37

(a) The figure is symmetrical (Figure 37(a)) and $\triangle ABC$ is equilateral. The tensions are equal (which can be confirmed by resolving horizontally).

$$2T\cos 30° = 10g \Rightarrow T = \frac{10g}{\sqrt{3}} = 56.6\text{ N}$$

C is $\frac{1}{2}\sqrt{3}$ m $= 86.6$ cm below AB.

(b) ABC is isosceles, and AC and BC are stretched by the same amount x.

Hooke's law $\Rightarrow T = \dfrac{\lambda x}{1} = 5gx \quad \uparrow\ 2T\cos\theta = 10g$

$$\Rightarrow 10gx\cos\theta = 10g \Rightarrow x\cos\theta = 1$$

From BCD, $\sin\theta = \dfrac{\frac{1}{2}}{1+x}$ and from above $x = 1/\cos\theta = \sec\theta$

$$\Rightarrow (1+x)\sin\theta = \tfrac{1}{2} \Rightarrow \left(1+\frac{1}{\cos\theta}\right)\sin\theta = \tfrac{1}{2} \Rightarrow \tan\theta + \sin\theta = \tfrac{1}{2}$$

Use Newton–Raphson method with the first approximation $\theta_1 = \sin^{-1}\frac{1}{4}$.

$$f(\theta) = \tan\theta + \sin\theta - \tfrac{1}{2} = 0 \Rightarrow \theta_2 = \theta_1 - \frac{f(\theta_1)}{f'(\theta_1)}$$

$$\Rightarrow \theta_2 = \theta_1 - \frac{\tan\theta_1 + \sin\theta_1 - \frac{1}{2}}{\sec^2\theta_1 + \cos\theta_1} = 0.2527 - \frac{0.2582 + 0.25 - \frac{1}{2}}{1.0667 + 0.9682}$$

$$= 0.2527 - \frac{0.0082}{2.0349}$$

$= 0.2527 - 0.0040 = 0.2487$ radians $= 14.25°$

$x = \sec\theta = 1.032$ and $x + 1 = \frac{1}{2}/\sin\theta = \frac{1}{2}/\sin 14.25° = 2.031$
$x = 1.032$ m $= 103.2$ cm and $T = 5gx = 50.57$ N

Example 12
A particle at the point $A(3, 2, 1)$ is acted on by forces $\mathbf{F}_1 = 2\mathbf{i} - \mathbf{j} + 3\mathbf{k}$ and $\mathbf{F}_2 = \mathbf{i} + \mathbf{j} - 2\mathbf{k}$ acting through that point. If the resultant acts through the point $B(6, y, z)$, find the values of y and z.

$\mathbf{F}_1 + \mathbf{F}_2 = 3\mathbf{i} + \mathbf{k}$. In terms of vectors $\mathbf{OB} = \mathbf{OA} + t(3\mathbf{i} + \mathbf{k})$
$\Rightarrow 6\mathbf{i} + y\mathbf{j} + z\mathbf{k} = 3\mathbf{i} + 2\mathbf{j} + \mathbf{k} + t(3\mathbf{i} + \mathbf{k})$
$\Rightarrow 6 = 3 + 3t;\ y = 2;\ z = 1 + t$
$\Rightarrow t = 1;\ z = 2.$

B is $(6, 2, 2)$, $y = 2$ and $z = 2$.

Chapter 15
Statics of a Rigid Body

Example 1
The centre of mass of a hollow cone of height h and base radius r is at a distance from the vertex of

A $\frac{1}{2}h$ **B** $\frac{1}{4}h$ **C** $\frac{2}{3}h$ **D** $\frac{3}{4}h$ **E** $\frac{1}{3}h$

Usually the position of the centre of mass (gravity) is given from the base of the cone. In the case of a solid cone it is $\frac{1}{4}h$ from the base. A hollow cone can be regarded as made up of many thin isosceles triangles, each of whose base is a small section of the circumference of the circular base of the cone and whose top vertex is the vertex of the cone. Each triangle has its centre of mass $\frac{1}{3}h$ from the base, i.e., $\frac{2}{3}h$ from the vertex of the cone. This result is confirmed in Example 10. ANSWER C

Example 2
The centre of mass of a lamina (flat plate) whose shape is bounded by the curve $y = f(x)$, the x- and y-axes and the line $x = a$ is $(\bar{x}, \bar{y})$ where

A $\bar{x} = \int_0^a xy\,dx$ and $\bar{y} = \int_0^a \frac{1}{2}y^2\,dx$

B $\bar{x} = \dfrac{\int_0^a xy\,dx}{\int_0^a y\,dx}$ and $\bar{y} = \dfrac{\int_0^a y^2\,dx}{\int_0^a y\,dx}$

C $\bar{x} = \dfrac{\int_0^a x^2y\,dx}{\int_0^a y\,dx}$ and $\bar{y} = \dfrac{\int_0^a xy^2\,dx}{\int_0^a y\,dx}$

D $\bar{x} = \dfrac{\int_0^a xy\,dx}{\int_0^a y\,dx}$ and $\bar{y} = \dfrac{\int_0^a \frac{1}{2}y^2\,dx}{\int_0^a y\,dx}$ **E** none of these.

The answer is justified in Example 9. Answer D is correct. The result is obtained by taking moments about each axis to find $\bar{x}$ and $\bar{y}$.

ANSWER D

Example 3
A weight of 4 N is suspended from an elastic string of length 6 m and modulus 12 N. The string is extended by

A 2 m **B** $2g$ m **C** 24 m **D** $1\frac{1}{2}$ m **E** 1 m

The tension in an elastic string of length a metres and modulus of elasticity λ which has been extended a further x metres is given by $T = \frac{\lambda}{a}x$. When the string is extended the tension in the string is balanced by the weight, so

$$T = 4 = \tfrac{12}{6}x \Rightarrow x = 2\,\text{m}$$

ANSWER A

Do not confuse weight with mass and insert $4g$ for T.

Multiple choice questions (Type B) Answer according to the table

A	B	C	D	E
1, 2, 3 correct	1, 3 only	2, 3 only	2 only	3 only

Example 4
The forces acting along the sides of the square $ABCD$

1 are in equilibrium **2** have resultant $\mathbf{i}-\mathbf{j}$
3 are equivalent to a force acting through the centre of the square.

The resultant of these forces has a horizontal component of $4\mathbf{i}-3\mathbf{i} = \mathbf{i}$ and a vertical component of $\mathbf{j}-2\mathbf{j} = -\mathbf{j}$. Therefore they cannot be in equilibrium.

Taking moments about the centre of the square gives a clockwise moment which is not zero, so the resultant will have a non-zero moment about the centre and will not act through the centre.

In fact, the resultant will act through (6, 6) if C is (2, 2).
Statement 2 only is correct.

ANSWER D

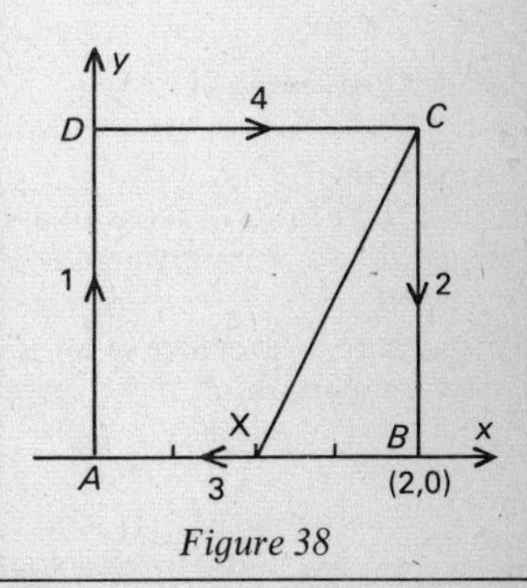

Figure 38

Example 5
In Figure 38 the resultant of the two forces, one along AD and the other along CB

1 has magnitude 3 **2** acts in the same direction as AD
3 has line of action given by $x = 4$

The resultant has magnitude 1 downwards through the point $(x, 0)$, where $1 \times x = 2 \times 2$ (moments about A) $\Rightarrow x = 4$.
Statement 3 only is true. ANSWER E

Example 6
In Figure 38, if $ABCD$ is a flat plate, the centre of mass G of the remaining shape when triangle BCX is removed lies on

1 $y = 2$ **2** $x = \frac{7}{9}$ **3** DB

If G is $(\bar{x}, \bar{y})$ then taking moments about AD gives

$$(\text{mass of square } ABCD) \times 1 = (\text{mass of } AXCDA) \times \bar{x} + (\text{mass of } BCX) \times 1\tfrac{2}{3}$$

The masses are proportional to the areas, so

$$4 \times 1 = 3 \times \bar{x} + 1 \times 1\tfrac{2}{3} \Rightarrow 3\bar{x} = 2\tfrac{1}{3} \Rightarrow \bar{x} = \tfrac{7}{9}$$

y is given by taking moments about AB so

$$4 \times 1 = 3 \times \bar{y} + 1 \times \tfrac{2}{3} \Rightarrow 3\bar{y} = 3\tfrac{1}{3} \Rightarrow \bar{y} = \tfrac{10}{9}$$

The equation of BD is $x + y = 2$ which is not satisfied by $G(\frac{7}{9}, \frac{10}{9})$. For G to lie on BD the centroid of BCX would have to lie on DB. This is not true; therefore, statement 2 only is correct.

ANSWER D

General questions

Example 7
Find the centre of mass of a square lamina $ABCD$ of side $2r$ when the square portion $ODEF$ is removed (Figure 39(a)).

By symmetry, the centre of mass lies on OE at G, where $GH = h$. Take moments about AH to give

$$(\text{mass of } ABCD) \times HO = (\text{mass of remainder}) \times HG + (\text{mass of } ODEF) \times HJ$$

Mass is proportional to area so replace ‘mass’ by ‘area’ to give

$$4r^2 \times r = (4r^2 - r^2) \times h + r^2 \times 1\tfrac{1}{2}r$$

$$4r^3 = 3r^2h + 1\tfrac{1}{2}r^3 \Rightarrow 3r^2h = 2\tfrac{1}{2}r^3 \Rightarrow h = \tfrac{5}{6}r$$

This type of question is solved by saying that the moments of the parts which make up the whole about any axis (or line in two dimensions) are equal to the moment of the whole body about that axis. If the figure were not symmetrical about HE it would be necessary to take moments about DC to find the distance of the centre of mass from that line.

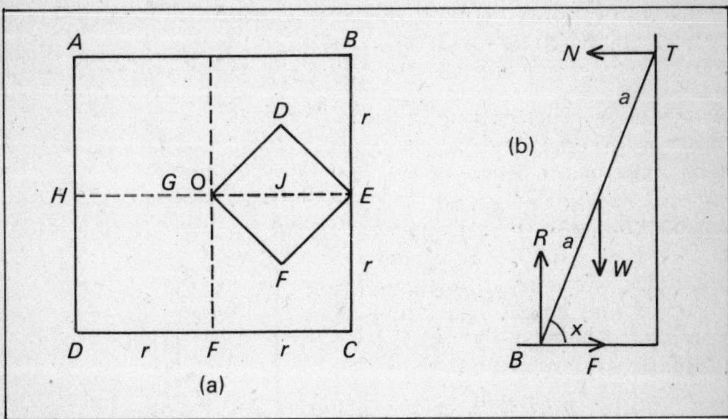

Figure 39

Example 8
The recommended angle for a ladder of weight W leaning against a wall is $x = \tan^{-1} 4$. With the ladder at this angle, standing on rough ground and leaning against a smooth wall, find the coefficient of friction between the ladder and the ground if the ladder is in limiting equilibrium. Find the value of μ which allows a man of weight $4W$ to climb up the ladder.

With the notation in Figure 39(b), $\leftarrow\ N = F \quad \uparrow\ R = W$

$$\overset{\frown}{B}\quad Wa\cos x = N \times 2a\sin x \Rightarrow W = 2N\tan x \Rightarrow N = \frac{W}{8}$$

$$F = N = \frac{W}{8} \Rightarrow = \frac{F}{R} = \frac{W/8}{W} = \frac{1}{8}$$

If the man climbs a distance y up the ladder $\leftarrow N = F \quad \uparrow R = 5W$

$\overset{\frown}{B}$ $Wa\cos x + 4Wy\cos x = N \times 2a\sin x \Rightarrow 2aN = W\cot x(a+4y)$

$$\mu = \frac{F}{R} = \frac{W\cot x}{2a \cdot 5W}(a+4y) = \frac{a+4y}{40a}$$

$0 \leqslant y \leqslant 2a \Rightarrow \frac{1}{40} \leqslant \mu \leqslant \frac{9}{40}$ and μ must be at least $\frac{9}{40}$ to enable the man to climb up the ladder.

> **Example 9**
> Find the position of the centre of mass of a semicircular lamina of radius a (Figure 40). If it is suspended from A find the angle AB makes with the vertical.

By symmetry, the centre of mass must lie on Ox, and $\bar{x}$, the x coordinate of the centre of mass of each quadrant AOX and BOX, must be the same. It is only necessary to find $\bar{x}$ for AOX. By first principles, consider a strip of width δx and height y distant x from AO, where $x^2+y^2 = a^2$. If the mass per unit area is m then

$$(\text{mass of } AOX) \times \bar{x} = \sum_{\delta x \to 0} my\delta x \times x = m\int_0^a xy\,dx$$

Since the total mass of AOX (in general) is $\int my\,dx$ we have a formula for $\bar{x}$ (in general) of

$$\bar{x} = \frac{\int xy\,dx}{\int y\,dx}$$

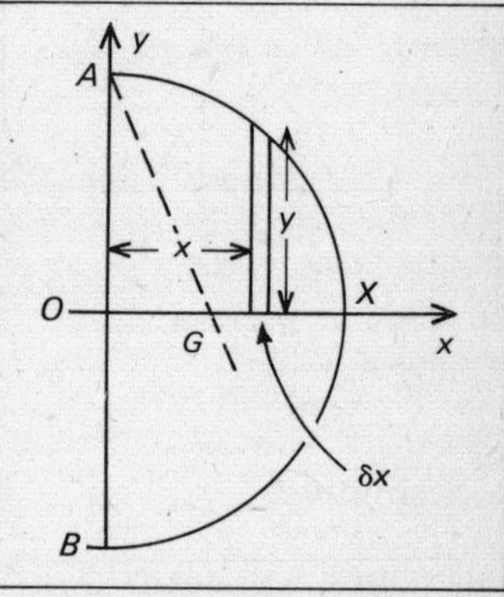

Figure 40

Many students remember this and quote the result for a lamina bounded by the curve $y = f(x)$. Taking moments about Ox for the same strip

$$(\text{mass of } AOX) \times \bar{y} = \Sigma my \cdot \delta x \cdot \tfrac{1}{2}y = m\int \tfrac{1}{2}y^2\,dx$$

since the centre of mass of each strip is half-way up the strip; hence the use of $\frac{1}{2}y$.

So $$\bar{y} = \frac{\int \frac{1}{2}y^2\,dx}{\int y\,dx}$$

In this example $\bar{x} = \dfrac{\int x\sqrt{(a^2-x^2)}\,dx}{\int y\,dx} = \dfrac{[-\frac{1}{3}(a^2-x^2)^{3/2}]_0^a}{\frac{1}{4}\pi a^2}$

$$= \frac{\frac{1}{3}a^3}{\frac{1}{4}\pi a^2} = \frac{4a}{3\pi}$$

As a check $\bar{y} = \dfrac{\int \frac{1}{2}(a^2-x^2)\,dx}{\frac{1}{4}\pi a^2} = \dfrac{[\frac{1}{2}(a^2x-\frac{1}{3}x^3)]_0^a}{\frac{1}{4}\pi a^2}$

$$= \frac{\frac{1}{2}\cdot\frac{2}{3}a^3}{\frac{1}{4}\pi a^2} = \frac{4a}{3\pi}$$

For a quadrant, $\bar{x}$ should be the same as $\bar{y}$ by symmetry (the centre of mass must lie on $y = x$) but calculation confirms this.

When suspended, AG is vertical so if AOB is inclined at $x°$ to the vertical

$$\tan x = \frac{OG}{AO} = \frac{4a/3\pi}{a} = \tfrac{4}{3}\pi = 0.4244 \Rightarrow x = 23°$$

Example 10
Find the centre of mass of a solid cone of height h and base radius r. Use this to find the centre of mass of a hollowed cone of external dimensions H and R and internal dimensions h and r. If these dimensions are in the same ratio, i.e., the interior and exterior cones have the same angle, deduce the centre of mass of a hollow cone.

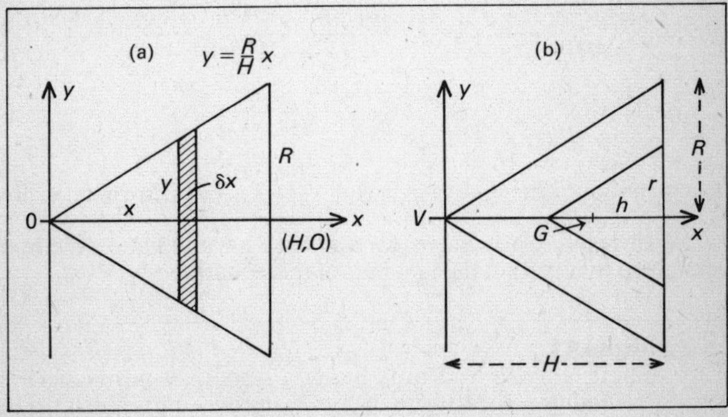

Figure 41

In Figure 41(a) consider a slice of width δx, distant x from vertex O. Its radius is $y = \frac{R}{H}x$ and if the mass per unit volume is m

$$\overset{\frown}{Oy} \quad (\text{mass of cone}) \times \bar{x} = \Sigma\,(\text{mass of slice}) \times x$$

$$= m\int \pi y^2 x\,dx = m\pi\int x \cdot \frac{R^2}{H^2}x^2\,dx$$

$$\tfrac{1}{3}\pi R^2 H\bar{x} = \pi\frac{R^2}{H^2}\left[\frac{x^4}{4}\right]_0^H = \pi\frac{R^2}{H^2}\frac{H^4}{4} \Rightarrow \bar{x} = \frac{3H}{4}$$

If the centre of mass of the hollowed cone (Figure 41(b)) is at G

$$\overset{\frown}{Vy} \quad \tfrac{1}{3}\pi R^2 H \cdot \frac{3H}{4} = \tfrac{1}{3}\pi r^2 h(H - \tfrac{1}{4}h) + \tfrac{1}{3}\pi(R^2 H - r^2 h)VG$$

$$\tfrac{3}{4}R^2H^2 = r^2hH - \tfrac{1}{4}r^2h^2 + (R^2H - r^2h)VG \quad \text{and} \quad \frac{R}{H} = \frac{r}{h} \Rightarrow r = \frac{hR}{H}$$

$$VG = \frac{3R^2H^2 - 4r^2hH + r^2h^2}{4(R^2H - r^2h)} = \frac{3R^2H^2 - 4hH \cdot \dfrac{h^2R^2}{H^2} + h^2 \cdot \dfrac{R^2h^2}{H^2}}{4\left(R^2H - h \cdot \dfrac{R^2h^2}{H^2}\right)}$$

using $r = \frac{hR}{H}$

$$VG = \frac{3H^4 - 4h^3H + h^4}{4(H^3 - h^3)} = \frac{(H-h)(3H^3 + 3H^2h + 3Hh^2 - h^3)}{4(H-h)(H^2 + Hh + h^2)}$$

$$= \frac{3H^3 + 3H^2h + 3Hh^2 - h^3}{4(H^2 + Hh + h^2)}$$

For a hollow cone $h \to H \Rightarrow VG = \frac{9H^3 - H^3}{12H^2} = \tfrac{2}{3}H$ which is the required result. This can be derived with an integration method similar to that used in the first part of this equation.

Example 11
A smooth uniform rod rests inside a smooth hemispherical bowl of radius r. Investigate the positions of equilibrium if the rod has length (a) r and (b) $3r$.

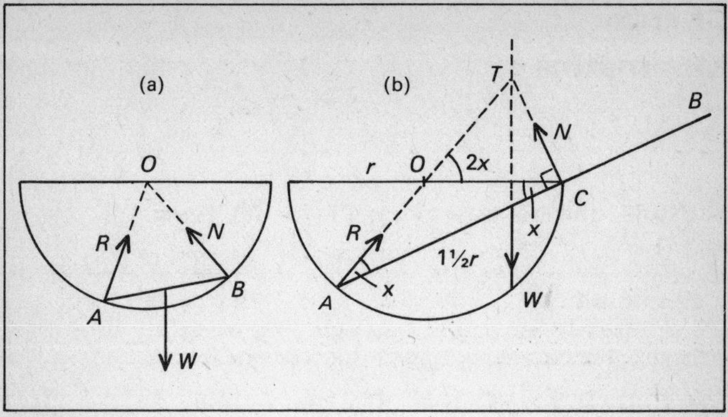

Figure 42

(a) Figure 42(a) shows the rod of length r inside the bowl at an angle. There are three forces acting on the rod: its weight and the reactions at each end, which must be perpendicular to the surfaces in contact and therefore act through the top centre of the bowl. If a rigid body is in equilibrium under the action of three forces, these three forces must be concurrent (act through the same point). The weight must then act through the centre O, so the centre of AB must be vertically below O which means that the rod will be horizontal.

(b) When the rod's length is $3r$ it can rest in equilibrium across the top of the bowl, but this is a trivial case.

Figure 42(b) shows a possible position with one end resting inside the bowl. R acts through the centre of the bowl O, but the reaction at C is perpendicular to the rod.

If $\angle OAC = x$ then $\angle OCA = x$ and $\angle COT = 2x$ (exterior angle).

$\circlearrowleft A \; N \cdot AC = W \cdot 1\tfrac{1}{2}r\cos x \quad \uparrow \; R\sin 2x + N\cos x = W$

$\leftarrow \; R\cos 2x = N\sin x$

$AC = 2r\cos x \Rightarrow N \cdot 2r\cos x = W \cdot 1\tfrac{1}{2}r\cos x$

$$\Rightarrow N = \tfrac{3}{4}W, \; R = \frac{3W\sin x}{4\cos 2x}$$

With these values of N and R, $R\sin 2x + N\cos x = W$ becomes

$$\frac{3W\sin x}{4\cos 2x} \cdot \sin 2x + \tfrac{3}{4}W\cos x = W$$

$\Rightarrow 3\sin x \cdot 2\sin x\cos x + 3\cos x \cdot \cos 2x = 4\cos 2x$

$\Rightarrow 6\cos x(1-\cos^2 x) + 3\cos x(2\cos^2 x - 1) = 4(2\cos^2 x - 1)$

$\Rightarrow 8\cos^2 x - 3\cos x - 4 = 0 \Rightarrow x = 23.2°$

Chapter 16
Kinematics

Multiple choice questions (Type A) Select the correct answer

Example 1
For a particle moving in a straight line, where $x = 3t^2 + 6t$, the distance covered during the third second is

A 95 m **B** 9 m **C** 24 m **D** 21 m **E** 18 m

The distance covered during the third second is the difference between x (for $t = 3$) and x (for $t = 2$).

$x_3 - x_2 = 27 + 18 - 12 - 12 = 21$ m ANSWER D

Example 2
A particle moves along Ox so that at time t its distance x is given by $x = 3 \sin 4t$. At time $t = \frac{\pi}{2}$ its acceleration is

A 48 **B** 12 **C** 0 **D** -12 **E** -48

Acceleration $a = \dfrac{d^2x}{dt^2} = -16 \times 3 \sin 4t = -48 \sin 4t = 0$ when $t = \frac{\pi}{2}$.

ANSWER C

Example 3
For a particle moving with uniform acceleration a, starting velocity u, whose velocity is v at time t and whose distance is s at time t

A $v = u - at$ **B** $s = vt - \frac{1}{2}at^2$ **C** $s = ut - \frac{1}{2}at^2$

D $v^2 + u^2 = as$ **E** $s = \dfrac{(u-v)t}{2}$

There are five formulae for uniform acceleration:

$$v = u + at,\ s = ut + \tfrac{1}{2}at^2,\ s = vt - \tfrac{1}{2}at^2,\ s = \frac{(u+v)t}{2},\ v^2 = u^2 + 2as$$

Statement B only is correct. ANSWER B

Example 4
A particle moving in a straight line covers a distance of 1 m in the first second and 2 m in the second second. Its acceleration is

A $\frac{1}{2}$ m s^{-2} **B** 1 m s^{-2} **C** $1\frac{1}{2}$ m s^{-2} **D** 2 m s^{-2}
E none of these

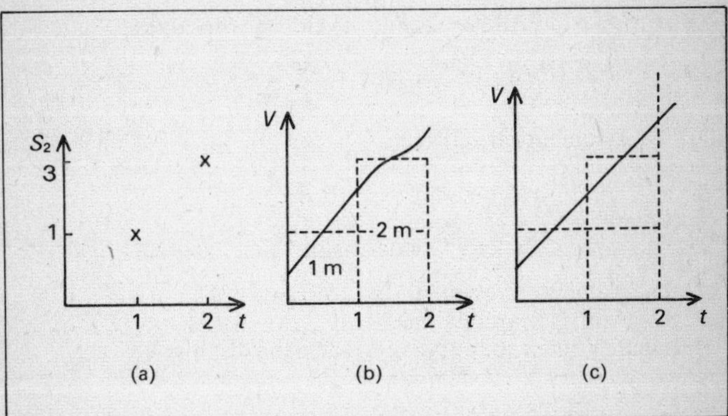

Figure 43

The information given is represented in Figure 43(a), with distance against time, and in Figure 43(b), with velocity against time. In this case the distance travelled is equal to the area under the graph.

If the acceleration is constant (Figure 43(c)) then B, 1 m s^{-2}, is the correct answer, but there is no guarantee of constant acceleration. Many graphs can be drawn (Figure 43(b) is an example) in which the areas are 1 m and 2 m so the correct answer is E.

ANSWER E

Multiple choice questions (Type B) Answer according to the table

A	B	C	D	E
1, 2, 3 correct	1, 3 only	2, 3 only	2 only	3 only

Example 5
The acceleration of a particle moving in a straight line whose displacement is x at a time t is

1 $\frac{d^2x}{dt^2}$ **2** $\frac{dv}{dt}$ **3** $v\frac{dv}{dx}$ where v is its velocity

v is the rate of change of displacement with time, i.e., $v = \frac{dx}{dt}$.
Acceleration is the rate of change of velocity with time so

$$a = \frac{dv}{dt} = \frac{d}{dt}\left(\frac{dx}{dt}\right) = \frac{d^2x}{dt^2} \text{ also } a = \frac{dv}{dx} \times \frac{dx}{dt} = \frac{dv}{dx} \times v = v\frac{dv}{dx}$$

All three statements are correct. ANSWER A

Example 6
For a graph of velocity (y-axis) against time (x-axis)

1 the gradient represents the acceleration
2 the graph is a straight line
3 the area under the curve represents the distance covered

For a two-dimensional graph of velocity against time the motion must be one-dimensional, i.e., in a straight line.

Statement 1 is correct since acceleration is rate of change of velocity.
Statement 2 is correct only if the acceleration is uniform (constant).
Statement 3 is correct if care is taken when the velocity is negative. Instead of distance, the term displacement would be better.

Statements 1 and 3 only are correct. ANSWER B

Example 7
Two particles A and B are thrown upwards vertically from ground level, A with velocity $10\,m\,s^{-1}$ and B with velocity $5\,m\,s^{-1}$ one second later. Taking g (acceleration due to gravity as $10\,m\,s^{-2}$).

1 A travels twice as high as B
2 their velocities when each hits the ground are in the ratio 2:1
3 B reaches the ground first

To find A's greatest height $v^2 = u^2 + 2as \Rightarrow 0 = 10^2 - 2gs$

$$\Rightarrow s = \frac{100}{20} = 5\,\text{m}.$$

This occurs at time t where $v = u + at \Rightarrow 0 = 10 - gt \Rightarrow t = 1$ second. A hits the ground with speed v given by

$$v^2 = u^2 + 2as \Rightarrow v^2 = 0 + 2g5 \Rightarrow v = 10$$

This occurs at time t where $v = u + at \Rightarrow 10 = 0 + gt \Rightarrow t = 1$ second from top. This confirms that downward flight is the reverse of upwards motion.

To find B's greatest height $v^2 = u^2 + 2as \Rightarrow 0 = 5^2 - 2gs$

$$\Rightarrow s = \frac{25}{20} = 1.25\,\text{m}.$$

This occurs at time t where $v = u + at \Rightarrow 0 = 5 - gt \Rightarrow t = \frac{1}{2}$ second. B hits the ground with velocity $5\,\text{m}\,\text{s}^{-1}$ after a time of 1 second. A and B hit the ground at the same time (B started 1 second later).

Statement 2 only is correct. ANSWER D

General questions

Example 8
A particle of unit mass starts at rest with position vector $\mathbf{i} + 3\mathbf{j}$ at $t = 0$ and is acted on by a constant force of $2\mathbf{i} + \mathbf{j}$ acting through $\mathbf{r} = \mathbf{i} + 3\mathbf{j}$. Find its velocity at time t and its position vector when $t = 3$.

Constant force $2\mathbf{i} + \mathbf{j} \Rightarrow$ constant acceleration of $2\mathbf{i} + \mathbf{j}$ (unit mass)

$\mathbf{v} = \mathbf{u} + \mathbf{a}t \Rightarrow \mathbf{v} = (2\mathbf{i} + \mathbf{j})t$

$\mathbf{r} = \mathbf{u}t + \frac{1}{2}\mathbf{a}t^2 = \frac{1}{2}(2\mathbf{i} + \mathbf{j})t^2 = 4\frac{1}{2}(2\mathbf{i} + \mathbf{j}) = 9\mathbf{i} + 4\frac{1}{2}\mathbf{j}$

Example 9
A train on a level track of mass 20 tonnes pulls 2 trucks A and B, each of mass 10 tonnes. The train provides a tractive force of 40 000 N. Find the acceleration of the train and the tensions in the tow-bars between the trucks and between the engine and A.

The equation of motion for the train is $F - T_1 = 20\,000a$. (1)
The equation of motion for the truck A is $T_1 - T_2 = 10\,000a$. (2)
The equation of motion for the truck B is $T_2 = 10\,000a$. (3)
Adding (1), (2) and (3) gives $F = 40\,000a = 40\,000 \Rightarrow a = 1\,\mathrm{m\,s^{-2}}$.
This is equivalent to regarding the train and trucks as a complete body of mass 40 tonnes.
$(3) \Rightarrow T_2 = 10\,000\,\mathrm{N}$
$(2) \Rightarrow T_1 = T_2 + 10\,000a = 20\,000\,\mathrm{N}$

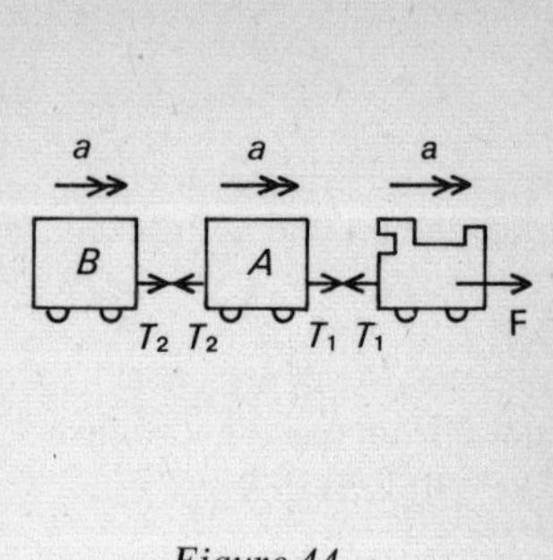

Figure 44

Example 10
A particle moves in a straight line so that its displacement x from the origin O at time t seconds is given by $x = t^3 - 6t^2 + 11t - 6$. Examine its motion.

x is a cubic in t with three possible zero values for x.

$$v = \frac{dx}{dt} = 3t^2 - 12t + 11 = 0 \text{ when } t = \frac{12 \pm \sqrt{144 - 4 \times 3 \times 11}}{6}$$

$$= \frac{12 \pm \sqrt{12}}{6} = \frac{12 \pm 2\sqrt{3}}{6} = \frac{6 \pm \sqrt{3}}{3} = 2 \pm 0.577$$

$= 2.577$ or 1.423 seconds.

$$a = \frac{dv}{dt} = 6t - 12 < 0 \text{ for } t < 2 \text{ and } a > 0 \text{ for } t > 2.$$

$t = 0 \Rightarrow x = -6$ and $v = 11$ so the particle is moving towards 0 from the negative side, but slowing down ($a < 0$ for $t < 2$).

At $t = 1.423$, $v = 0$, the particle stops momentarily and moves backwards until $t = 2.577$ ($v = 0$) and then moves forwards again.

In fact, $x = (t-1)(t-2)(t-3)$ which means that the particle passes through 0 at times 1, 2 and 3 seconds.

Example 11
A particle moving in a straight line with uniform (constant) acceleration covers distances of 20 m and 50 m in successive seconds. Find its acceleration and its speed at the start of the first second.

With the usual notation (see Example 3) the unknowns which we have to find are u (starting speed) and a (acceleration) and it is better to try and express the problem in terms of these variables.

For the first second $v = u + at \quad \Rightarrow v = u + a$

$$s = ut + \tfrac{1}{2}at^2 \Rightarrow 20 = u + \tfrac{1}{2}a \qquad (1)$$

For the second second U (starting speed).

$$U = u + a \qquad \Rightarrow V = u + 2a$$

$$S = Ut + \tfrac{1}{2}at^2 \Rightarrow 50 = u + a + \tfrac{1}{2}a = u + 1\tfrac{1}{2}a \qquad (2)$$

$(2) - (1) \Rightarrow 30 = a \Rightarrow u = 5$ so $u = 5\,\text{m s}^{-1}$ and $a = 30\,\text{m s}^{-2}$.

> **Example 12**
> How long does it take for the minute hand of a clock to 'catch up' again with the hour hand.

Both hands are together at 12 o'clock and again at roughly 6 minutes past one.

The angular velocity of the minute hand is $\frac{360}{60}$ degrees per minute.
The angular velocity of the hour hand is $\frac{30}{60}$ degrees per minute.
After t minutes the hour hand has travelled through an angle of $\frac{1}{2}t°$.
After t minutes the minute hand has travelled through an angle of $6t°$.
When they coincide again the minute hand has covered an extra 360°,
so $6t = \frac{1}{2}t + 360 \Rightarrow 5.5t = 360 \Rightarrow t = 65.\dot{4}\dot{5}$ minutes
$= 65$ minutes $27.\dot{2}\dot{7}$ seconds

The easier way to solve this problem is to remember that the hands coincide 11 times in 12 hours, so the time between each is $\frac{12}{11}$ hours $= 65.\dot{4}\dot{5}$ minutes!

> **Example 13**
> The position vector of a particle of mass 5 kg at time t is given by $\mathbf{r} = \frac{1}{2}t^2\mathbf{i} + \frac{1}{3}t^2\mathbf{j} + \frac{1}{4}t^4\mathbf{k}$. Find the force acting on the particle at time $t = 2$.

Force is given by $\mathbf{F} = m\mathbf{a}$ and $\mathbf{a} = \dfrac{d^2\mathbf{r}}{dt}$.

$$\mathbf{v} = \frac{d\mathbf{r}}{dt} = t\mathbf{i} + t^2\mathbf{j} + t^3\mathbf{k} \Rightarrow$$

$$\mathbf{a} = \frac{d\mathbf{v}}{dt} = \frac{d^2\mathbf{r}}{dt} = \mathbf{i} + 2t\mathbf{j} + 3t^2\mathbf{k}$$

$$t = 2 \Rightarrow \mathbf{a} = \mathbf{i} + 4\mathbf{j} + 12\mathbf{k} \Rightarrow \mathbf{F} = m\mathbf{a} = 5\mathbf{a} = 5\mathbf{i} + 20\mathbf{j} + 60\mathbf{k}$$

whose magnitude $= 5\sqrt{1^2 + 4^2 + 12^2} = 5\sqrt{161} = 63.44$ N.

Example 14
A particle moving in a straight line from rest has acceleration at time t of $t + t^2$. Find the distance covered in the first 2 seconds.

$$a = \frac{dv}{dt} = t + t^2 \Rightarrow v = \int (t + t^2)\,dt = \tfrac{1}{2}t^2 + \tfrac{1}{3}t^3 + k$$

$$t = 0, v = 0 \Rightarrow k = 0 \text{ so } v = \frac{dx}{dt} = \tfrac{1}{2}t^2 + \tfrac{1}{3}t^3$$

$$\Rightarrow x = \int_0^2 (\tfrac{1}{2}t^2 + \tfrac{1}{3}t^3)\,dt = [\tfrac{1}{6}t^3 + \tfrac{1}{12}t^4]_0^2 = \tfrac{8}{6} + \tfrac{16}{12} - 0 = \tfrac{4}{3} + \tfrac{4}{3} = 2\tfrac{2}{3}\text{ m}$$

Example 15
Examine the motion of a particle moving in a plane so that its position at time t is given by $\mathbf{r} = x\mathbf{i} + y\mathbf{j} = r\cos\omega t\mathbf{i} + r\sin\omega t\mathbf{j}$ where r and ω are constant.

$x = r\cos\omega t, y = r\sin\omega t \Rightarrow x^2 + y^2 = r^2(\cos^2\omega t + \sin^2\omega t) = r^2$.
This is the equation of a circle centre the origin and radius r.

$$\dot{x} = \frac{dx}{dt} = -r\omega\sin\omega t \qquad \dot{y} = \frac{dy}{dt} = r\omega\cos\omega t$$

$$\Rightarrow \text{velocity } \mathbf{v} = \frac{d\mathbf{r}}{dt} = \dot{x}\mathbf{i} + \dot{y}\mathbf{j} = -r\omega\sin\omega t\mathbf{i} + r\omega\cos\omega t\mathbf{j}$$

$\mathbf{v} \times \mathbf{r} = -r^2\omega\cos\omega t\sin\omega t + r^2\omega\sin\omega t\cos\omega t = 0 \Rightarrow v \perp$ radius vector, i.e., the velocity is directed along the tangent.

$$v = |\mathbf{v}| = \sqrt{r^2\omega^2(\sin^2\omega t + \cos^2\omega t)} = r\omega$$

The particle completes a circle when $\cos\omega t$ and $\sin\omega t$ complete one cycle of values, i.e., when ωt changes by 2π. ω is the change in angle per unit time, i.e., the angular velocity in radians/second.

$$\text{Acceleration } \mathbf{a} = \frac{d\mathbf{v}}{dt} = \frac{d^2\mathbf{r}}{dt^2} = -r\omega^2\cos\omega t\mathbf{i} - r\omega^2\sin\omega t\mathbf{j} = -\omega^2\mathbf{r}$$

which shows that the acceleration has magnitude $r\omega^2$ and is directed towards the centre of the circle.

Example 16
For a particle moving in a straight line with initial velocity $5\,\mathrm{m\,s^{-1}}$ find expressions for the velocity, v, and the displacement, x, from 0, the starting position, when the acceleration $a = -9x$.

$a = \dfrac{dv}{dt} = -9x$ which involves three variables v, t and x.

$a = \dfrac{d^2x}{dt^2} = -9x$ which is not an easy differential equation to solve.

$a = v\dfrac{dv}{dx} = -9x$ and provides the best starting place.

$$v\frac{dv}{dx} = -9x \Rightarrow \int v\,dv = \int -9x\,dx \Rightarrow \tfrac{1}{2}v^2 = -4\tfrac{1}{2}x^2 + k$$

$$x = 0, v = 5 \Rightarrow 12\tfrac{1}{2} = -4\tfrac{1}{2}\times 0 + k \Rightarrow k = 12\tfrac{1}{2}$$

$$v^2 = -9x^2 + 25 \Rightarrow \left(\frac{dx}{dt}\right)^2 = 25 - 9x^2 \Rightarrow \frac{dx}{dt} = \sqrt{25-9x^2}$$

$$\Rightarrow t = \int \frac{1}{\sqrt{25-9x^2}}\cdot dx = \tfrac{1}{3}\sin^{-1}(\tfrac{3}{5}x) + c;\ t = 0, x = 0 \Rightarrow 0 = 0 + c$$

$$\Rightarrow 3t = \sin^{-1}(\tfrac{3}{5}x) \Rightarrow \tfrac{3}{5}x = \sin 3t \Rightarrow x = \tfrac{5}{3}\sin 3t$$

This is an example of SHM (Simple Harmonic Motion) in which the particle oscillates about a mean position in the form of a sine curve.

Example 17
A particle travelling in a straight line has acceleration inversely proportional to its velocity. Initially ($x = 0$) its velocity is $4\,\mathrm{m\,s^{-1}}$ and its acceleration is $2\,\mathrm{m\,s^{-2}}$. Find its displacement, velocity and acceleration at time $t = 3\,\mathrm{s}$.

$a \propto \dfrac{1}{v} \Rightarrow a = \dfrac{k}{v} \Rightarrow k = av = 8$ from initial conditions.

$$a = \frac{dv}{dt} = \frac{k}{v} \Rightarrow \int v\,dv = \int 8\,dt \Rightarrow \tfrac{1}{2}v^2 = 8t + c$$

$$t = 0, v = 4 \Rightarrow 8 = c \Rightarrow \tfrac{1}{2}v^2 = 8t + 8 \Rightarrow v^2 = 16t + 16 = \left(\frac{dx}{dt}\right)^2$$

$$\Rightarrow \frac{dx}{dt} = \sqrt{16t+16} \Rightarrow x = \int \sqrt{(16t+16)}\, dt = \tfrac{1}{16} \frac{(16t+16)^{3/2}}{\frac{3}{2}} + k$$

$$x = 0, t = 0 \Rightarrow 0 = \frac{2}{3} \times \frac{64}{16} + k \Rightarrow k = -2\tfrac{2}{3}.$$

$$x = \tfrac{1}{24}(16t+16)^{3/2} - 2\tfrac{2}{3}$$

$$t = 3 \Rightarrow x = \frac{(48+16)^{3/2}}{24} - 2\tfrac{2}{3} = \frac{8^3}{24} - \frac{8}{3} = \frac{64}{3} - \frac{8}{3} = 18\tfrac{2}{3}\,\text{m}$$

$$v = \sqrt{16t+16} = \sqrt{64} = 8\,\text{m}\,\text{s}^{-1}$$

$$a = \frac{k}{v} = \frac{8}{8}\,\text{m}\,\text{s}^{-2} = 1\,\text{m}\,\text{s}^{-2}$$

Example 18
A particle accelerates for 5 m, then maintains the speed acquired for 20 m and then decelerates for 10 m. If the total time taken is 10 s, find the acceleration, speed attained and the deceleration.

Let the particle accelerate for a time t to gain a speed of v with acceleration a. The graph of speed against time is shown in Figure 45.

$$v = 0 + at \quad \text{and} \quad 5 = \tfrac{1}{2}vt = \tfrac{1}{2}at^2$$

For the period of deceleration,

$$v = dT \quad \text{and} \quad 10 = \tfrac{1}{2}vT = \tfrac{1}{2}dT^2$$

where d is the deceleration and T the time.

$$5 = \tfrac{1}{2}vt \quad \text{and} \quad 10 = \tfrac{1}{2}vT \Rightarrow T = 2t$$

The time taken for the particle to cover the middle section at constant speed v is $10-t-T = 10-3t$.

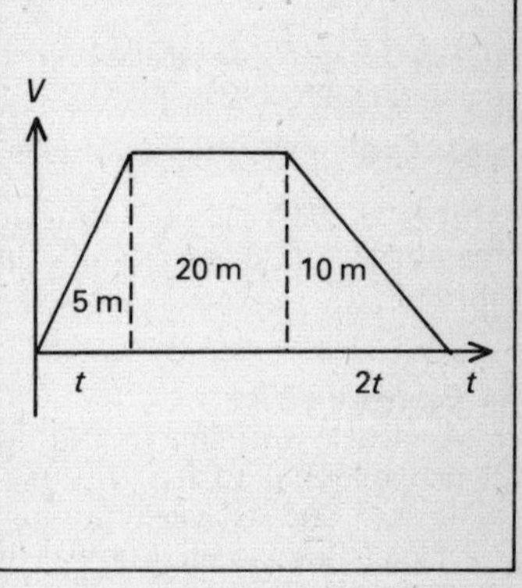

Figure 45

The distance covered is $20\,\text{m} \Rightarrow (10-3t)v = 20$
$10v - 3tv = 20 \Rightarrow 10v - 30 = 20$ since $tv = 10$.
$10v = 50 \Rightarrow v = 5\,\text{m}\,\text{s}^{-1} \Rightarrow t = 2$ and $T = 2t = 4\,\text{s}$.
$at = v \Rightarrow a = \frac{5}{2} = 2\frac{1}{2}\,\text{m}\,\text{s}^{-2}$; $dT = v \Rightarrow d = \frac{5}{4} = 1\frac{1}{4}\,\text{m}\,\text{s}^{-2}$.

Chapter 17
Laws of Motion

Newtonian mechanics is based upon the three laws of motion formulated by Sir Isaac Newton. These laws form the basis of all the questions set involving the motion of a particle subject to a given number of forces.

These laws are:
(1) Every body remains at rest or in uniform motion unless acted upon by some external force.
(2) The rate of change of momentum of a moving body is proportional to the applied force and takes place in the direction of that force.
(3) To every action there is an equal and opposite reaction.

If a force $\mathbf{F}$ is applied to a particle of mass m the second law, which features in most calculations, can be expressed in the form

$$\frac{d}{dt}(m\mathbf{v}) = \mathbf{F} \qquad \text{where } \mathbf{v} \text{ is the velocity.}$$

The units are chosen so that the constant of proportionality is equal to unity.

If the mass is constant this becomes $\mathbf{F} = m\mathbf{a}$ where $\mathbf{a}$ is the acceleration.

The nature of the force may vary and although most questions only use constant forces some syllabuses include variable forces in such topics as resisted motion under gravity or simple harmonic motion.

Multiple choice questions (Type A) Select the correct answer

> **Example 1**
> A particle of mass 2 kg starts from rest at time $t = 0$ and moves under the action of a force $\mathbf{F} = 3\mathbf{i} + 2\mathbf{j} - \mathbf{k}$ newtons. The magnitude of its velocity (in m s^{-1}) after 2 seconds is
>
> **A** $2\sqrt{3}$ **B** $\sqrt{14}$ **C** $2\sqrt{14}$ **D** $\sqrt{3}$ **E** none of these

Using Newton's second law, $\mathbf{F} = m\mathbf{a}$, we have the

$$\text{acceleration given by, } \mathbf{a} = \tfrac{3}{2}\mathbf{i} + \mathbf{j} - \tfrac{1}{2}\mathbf{k}$$

Since this is constant, the velocity $\mathbf{v}$ is given by $\mathbf{v} = \mathbf{u} + \mathbf{a}t$.

In this case $\mathbf{u} = 0$ and hence $\mathbf{v} = \frac{3}{2}t\mathbf{i} + t\mathbf{j} - \frac{1}{2}t\mathbf{k}$.
If $t = 2$ then $\mathbf{v} = 3\mathbf{i} + 2\mathbf{j} - \mathbf{k}$.
The magnitude of the velocity $\mathbf{v} = \sqrt{3^2 + 2^2 + (-1)^2} = \sqrt{14}\,\mathrm{m\,s^{-1}}$.

ANSWER B

If the force is variable the method has to be slightly adapted and requires the use of calculus.

Example 2
A particle of mass 3 kg starts from rest from the point whose position vector is $2\mathbf{i} + \mathbf{j} + \mathbf{k}$ at time $t = 0$ and moves under the action of a force $\mathbf{F} = 3\mathbf{i} - 6t\mathbf{j} + 3t^2\mathbf{k}$. The position vector of the particle after 2 seconds is

A $4\mathbf{i} - \frac{5}{3}\mathbf{j} + \frac{7}{3}\mathbf{k}$ **B** $2\mathbf{i} - \frac{8}{3}\mathbf{j} + \frac{4}{3}\mathbf{k}$ **C** $8\mathbf{i} - 7\mathbf{j} + 4\mathbf{k}$
D $2\mathbf{i} - 4\mathbf{j} + \frac{8}{3}\mathbf{k}$ **E** none of these

Using $\mathbf{F} = m\mathbf{a}$ we have acceleration, $\mathbf{a} = \frac{1}{3}(3\mathbf{i} - 6t\mathbf{j} + 3t^2\mathbf{k})$

$$\therefore \qquad \text{acceleration, } \frac{d\mathbf{v}}{dt} = \mathbf{i} - 2t\mathbf{j} + t^2\mathbf{k}$$

Integrating with respect to t gives $\mathbf{v} = t\mathbf{i} - t^2\mathbf{j} + \frac{1}{3}t^3\mathbf{k} + \mathbf{C}$
Now $\mathbf{v} = 0$ when $t = 0 \Rightarrow \mathbf{C} = 0$.
Hence velocity $\mathbf{v} = \dfrac{d\mathbf{r}}{dt} = t\mathbf{i} - t^2\mathbf{j} + \frac{1}{3}t^3\mathbf{k}$.
Integrating again with respect to t gives

$$\mathbf{r} = \tfrac{1}{2}t^2\mathbf{i} - \tfrac{1}{3}t^3\mathbf{j} + \tfrac{1}{12}t^4\mathbf{k} + \mathbf{B}$$

Now $\mathbf{r} = 2\mathbf{i} + \mathbf{j} + \mathbf{k}$ when $t = 0 \Rightarrow \mathbf{B} = 2\mathbf{i} + \mathbf{j} + \mathbf{k}$.
The position vector is

$$\mathbf{r} = (\tfrac{1}{2}t^2 + 2)\mathbf{i} + (1 - \tfrac{1}{3}t^3)\mathbf{j} + (\tfrac{1}{12}t^4 + 1)\mathbf{k}$$

Hence when $t = 2$, $\qquad \mathbf{r} = 4\mathbf{i} - \frac{5}{3}\mathbf{j} + \frac{7}{3}\mathbf{k}$

ANSWER A

Note that answer B would be obtained if the constant of integration, B, is forgotten and answer C if the student forgets to divide by the mass. Answer D gives the velocity.

Example 3
A man of mass 80 kg stands in a lift which is ascending with a constant acceleration of $3\,\mathrm{m\,s^{-2}}$. If g is taken as $10\,\mathrm{m\,s^{-2}}$ the reaction between the man and the floor is

A 240 N **B** 560 N **C** 1040 N **D** 2400 N **E** none of these

Let the reaction between the man and the floor be R. The man's weight is $80g$ N and the forces are shown in Figure 46.

The effective upward force applied to the man is

$$R - 80g$$

Using $F = ma$ in the upward direction we have

$$R - 80g = 80 \times 3$$

$$\therefore \quad R = 240 + 800$$

$$= 1040 \text{ N}$$

The reaction is 1040 N.

ANSWER C

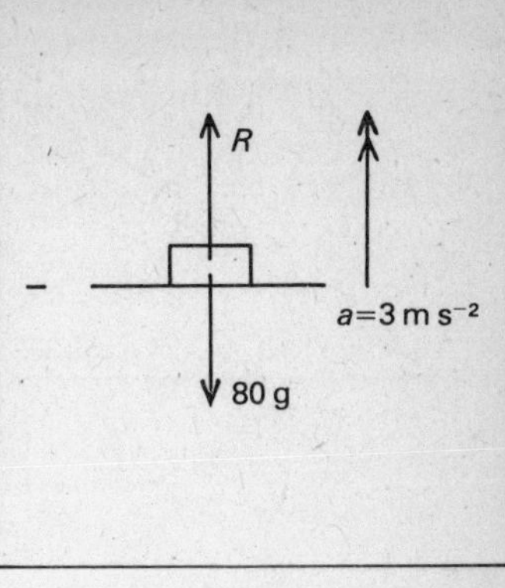

Figure 46

Example 4
A particle of mass 4 kg moves up the line of greatest slope of a rough plane inclined at 30° to the horizontal under the action of a constant force P_1 which acts along this line of greatest slope. If a constant force P_2, which acts down the line of greatest slope, is applied to the particle instead, it moves with an acceleration of 2 m s^{-2}. If the coefficient of friction, μ, is $1/\sqrt{3}$ and $g = 10 \text{ m s}^{-2}$, the ratio P_1/P_2 is

A $\frac{1}{2}$ **B** $\frac{5}{6}$ **C** $\frac{10}{3}$ **D** $\frac{5}{1}$ **E** none of these

The forces acting on the particle in each case are shown in Figure 47.

Note that the amount of dynamical friction is the same in both cases although it acts in opposite directions and always opposes the motion, i.e.

$$F = \mu R = 4\mu g \cos 30° = \frac{4g}{\sqrt{3}} \times \frac{\sqrt{3}}{2} = 2g$$

In case (a), the effective force in the direction of motion is $P_1 - F - 4g \sin 30°$.

Using $F = ma$ we have $P_1 - F - 4g \sin 30° = 0$

$$\therefore \quad P_1 - 2g - 2g = 0 \Rightarrow P_1 = 4g \text{ N}$$

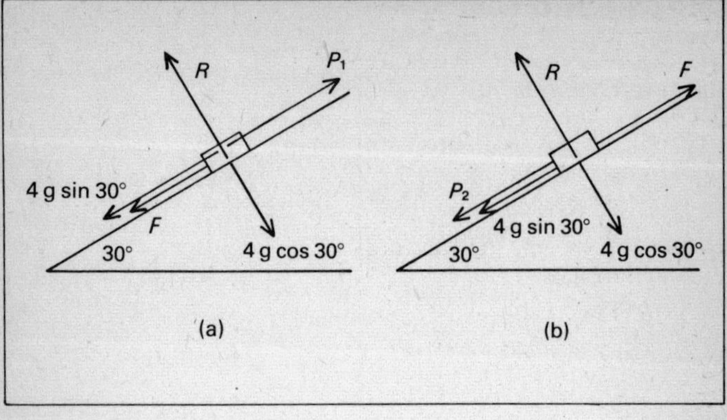

Figure 47

In case (b), the effective force down the plane is $P_2 - F + 4g \sin 30°$.

Using Newton's second law

$$P_2 - F + 4g \sin 30° = 4 \times 2$$

Hence $\quad P_2 - 2g + 2g = 8 \Rightarrow P_2 = 8\,\text{N}$

The ratio $\quad P_1/P_2 = 4g/8 = 5/1$

ANSWER D

General questions

Example 5
A particle of mass 3 kg is initially at rest at the point whose position vector is $\mathbf{i} - 3\mathbf{j} + 4\mathbf{k}$. It is then acted on by two forces $\mathbf{F}_1 = 2\mathbf{i} + 2\mathbf{j} + \mathbf{k}$ and $\mathbf{F}_2 = \mathbf{i} + 4\mathbf{j} - 4\mathbf{k}$. Find the acceleration of the particle and its position vector after 4 seconds.

The resultant force $\mathbf{F} = \mathbf{F}_1 + \mathbf{F}_2$
$= (2\mathbf{i} + 2\mathbf{j} + \mathbf{k}) + (\mathbf{i} + 4\mathbf{j} - 4\mathbf{k}) = 3\mathbf{i} + 6\mathbf{j} - 3\mathbf{k}$.

Using Newton's second law,

acceleration of the particle is $\frac{1}{3}(3\mathbf{i} + 6\mathbf{j} - 3\mathbf{k}) = \mathbf{i} + 2\mathbf{j} - \mathbf{k}$

Since this is constant, and $\mathbf{u} = 0$, then using $\mathbf{v} = \mathbf{u} + \mathbf{a}t$ the velocity is given by $\mathbf{v} = t(\mathbf{i} + 2\mathbf{j} - \mathbf{k})$.

Integrating with respect to t

$$\mathbf{r} = \tfrac{1}{2}t^2(\mathbf{i}+2\mathbf{j}-\mathbf{k})+\mathbf{C}$$

Now $\mathbf{r} = \mathbf{i}-3\mathbf{j}+4\mathbf{k}$ when $t = 0 \Rightarrow \mathbf{C} = \mathbf{i}-3\mathbf{j}+4\mathbf{k}$.

Hence
$$\mathbf{r} = \tfrac{1}{2}t^2(\mathbf{i}+2\mathbf{j}-\mathbf{k})+\mathbf{i}-3\mathbf{j}+4\mathbf{k}$$
$$\mathbf{r} = (1+\tfrac{1}{2}t^2)\mathbf{i}+(t^2-3)\mathbf{j}+(4-\tfrac{1}{2}t^2)\mathbf{k}$$

When $t = 4$, its position vector is $\mathbf{r} = 9\mathbf{i}+13\mathbf{j}-4\mathbf{k}$.

Example 6

A particle of mass 10 kg rests on a rough horizontal table. If the coefficient of friction is $\frac{1}{2}$, find the acceleration of the particle when a force equal to the weight of 8 kg is applied to the particle at an angle of 30° to the table. Take $g = 10\,\text{m s}^{-2}$.

The appropriate forces are shown in Figure 48.

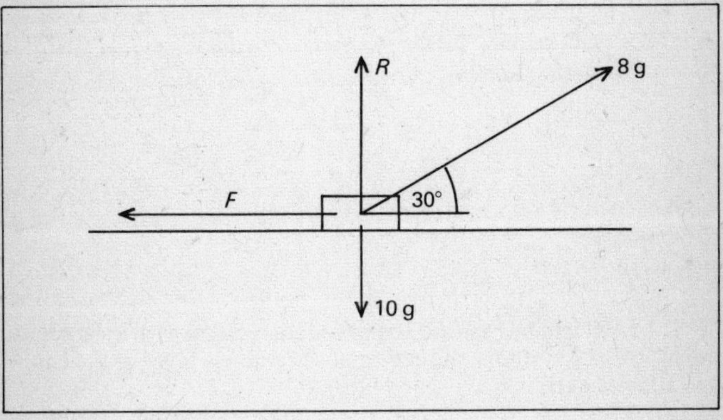

Figure 48

The applied force is $8g$ N. Let the acceleration of the particle be $a\,\text{m s}^{-2}$ along the plane of the table.
Since there is no motion vertically.

$$R+8g\sin 30° = 10g \Rightarrow R = 6g\text{ N}$$

Now as $\mu = \frac{1}{2}$ and $F = \mu R \Rightarrow F = \frac{1}{2}\times 6g = 3g$.

Applying Newton's second law horizontally,

$$8g\cos 30°-F = 10a$$

$$\Rightarrow 10a = 4\sqrt{3}g-3g \Rightarrow a = 4\sqrt{3}-3 = 3.93$$

Thus the acceleration of the particle is $3.93\,\text{m s}^{-2}$.

Example 7
A train consists of an engine of mass 100 tonnes and 10 coaches each of mass 15 tonnes. It is moving along a straight horizontal track with an acceleration of $0.2\,\mathrm{m\,s^{-2}}$ against resistances of 100 N/tonne. Find (a) the pull of the engine and (b) the tension in the coupling between the engine and the first coach if the resistances are divided between the engine and the coaches in the ratio of their masses.

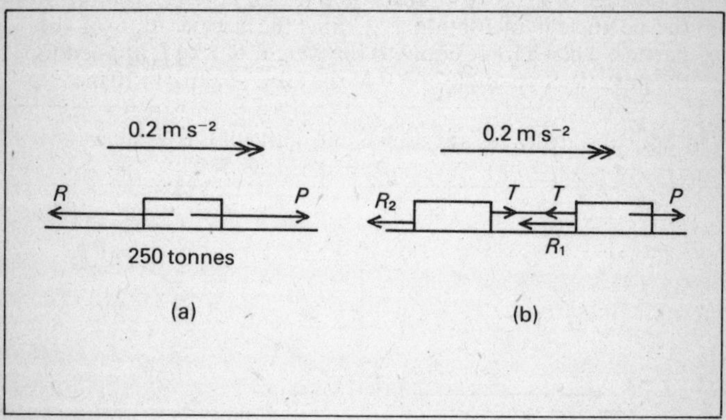

Figure 49

(a) The train can be regarded as one mass moving with an acceleration of $0.2\,\mathrm{m\,s^{-2}}$ under the action of the tractive force of the engine and the resistances (see Figure 49(a)).

Now the total resistances are 25 000 N. If P is the pull of the engine then, using $F = ma$,

$$P - 25\,000 = 250 \times 1000 \times \tfrac{1}{5} \Rightarrow P = 75\,000\,\mathrm{N}$$

Thus the pull of the engine is 75 kN.

(b) The forces are shown in Figure 49(b). Let T be the tension in the coupling and R_1 and R_2 be the resistances of the engine and the coaches respectively.
Since $R_1/R_2 = 100/150 \Rightarrow R_1 = \frac{2}{3}R_2$ or $R_1 = \frac{2}{5}R$ and $R_2 = \frac{3}{5}R$ where R is the total resistance.
Thus $R_1 = \frac{2}{5} \times 25\,000 = 10\,000\,\mathrm{N}$ and $R_2 = 15\,000\,\mathrm{N}$.

Consider the motion of the coaches which move under the action of the forces T and R_2.

Using Newton's second law,

$$T-R_2 = 150\times 1000\times \tfrac{1}{5} \Rightarrow T-15\,000 = 30\,000$$
$$\Rightarrow T = 45\,000\,\text{N}.$$

The tension in the coupling is 45 kN.

Note that in part (b) it is possible to consider the motion of the engine, i.e.

$$P-T-R_1 = 100\times 1000\times \tfrac{1}{5} \text{ giving } T = 45\,000\,\text{N}.$$

The problem of motion in a resisting medium involves variable forces and hence calculus methods are required.

Example 8

A particle of mass m is projected vertically upwards with a speed of $u\,\text{m}\,\text{s}^{-1}$ against a resistance of mkv^2 where k is a constant and v is its speed at time t seconds. If $u^2 = g/k$ find (a) the time taken to reach the highest point and (b) the greatest height.

(a) The total force on the particle is $mg+mkv^2$ downwards. Using Newton's second law

$$m\frac{dv}{dt} = -mg-mkv^2 \Rightarrow \frac{dv}{dt} = -(g+kv^2)$$

Hence $$\frac{1}{(g+kv^2)}\frac{dv}{dt} = -1 \Rightarrow \int_u^0 \frac{dv}{(g+kv^2)} = -\int_0^t dt$$

Thus $$\left[\frac{1}{\sqrt{kg}}\tan^{-1} v\sqrt{\frac{k}{g}}\right]_u^0 = [-t]_0^t \Rightarrow t = \frac{1}{\sqrt{kg}}\tan^{-1} u\sqrt{\frac{k}{g}}$$

Now since $u^2 = \dfrac{g}{k}$, $\quad u = \sqrt{\dfrac{g}{k}}$ and $\sqrt{k}u = \sqrt{g}$

Therefore, the time to reach the highest point is $t = \dfrac{1}{\sqrt{kg}}\tan^{-1} 1$.

i.e. $$t = \frac{\pi}{4}\times\frac{1}{\sqrt{kg}} \text{ or } t = \frac{u}{g}\times\frac{\pi}{4} = \frac{\pi u}{4g}$$

(b) In this part we need the relation between velocity and displacement.

Hence using $\dfrac{dv}{dt} = v\dfrac{dv}{ds}$ we have

$$mv\frac{dv}{ds} = -mg-mkv^2 \Rightarrow v\frac{dv}{ds} = -(g+kv^2)$$

Rearranging and integrating with respect to s as a definite integral (note that $v=u$ when $s=0$, and $v=0$ when $s=h$ (the greatest height)),

$$\therefore \qquad \int_u^0 \frac{v\,dv}{g+kv^2} = -\int_0^h ds$$

Thus $\quad [-s]_0^h = \left[\frac{1}{2k}\ln(g+kv^2)\right]_u^0 \Rightarrow -h = \frac{1}{k}\ln\left[\frac{g}{(g+ku^2)}\right]$

$$\therefore \qquad h = \frac{1}{2k}\ln\left[\frac{(g+ku^2)}{g}\right]$$

Since $u^2 = \frac{g}{k}$, the greatest height is $\frac{1}{2k}\ln 2$.

When two or more particles are connected by a light inextensible string, Newton's second law can be applied to the various masses.

Example 9
Masses of 7 kg and 3 kg are connected by a light inextensible string passing over a fixed smooth pulley. Find the acceleration of the system when it is released from rest. If, after falling 1.96 m, the 7 kg mass strikes an inelastic table find the further time before the 3 kg mass is at rest instantaneously assuming that it does not reach the pulley during the motion. Take the value of g as 9.8 m s^{-2}.

Let the tension in the string be T N and the acceleration of the system be a m s^{-2}. Figure 50 shows the initial position of the masses, the position when the 7 kg mass strikes the table and the subsequent situation where the string is slack.

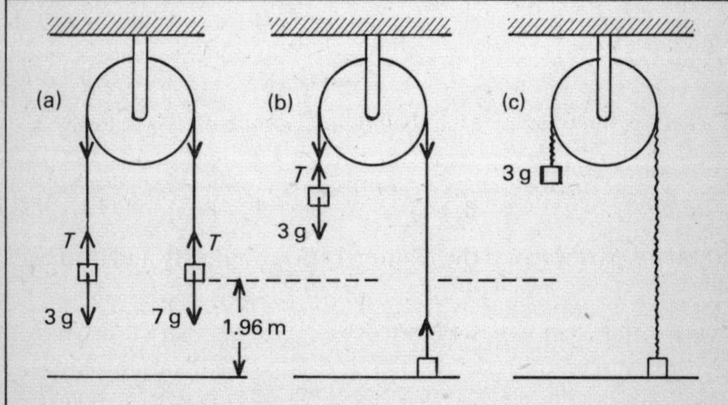

Figure 50

Using Newton's second law the equations of motion are,

on the 7 kg mass: $7g - T = 7a$ (1)

on the 3 kg mass: $T - 3g = 3a$ (2)

Adding (1) and (2) gives $4g = 10a \Rightarrow a = \frac{2}{5}g$

Hence, if v is the velocity of the particles after travelling 1.96 m we have, using $v^2 = u^2 + 2as$

$$v^2 = 0 + 2(\tfrac{2}{5}g)1.96 \Rightarrow v = 3.92\,\text{m}\,\text{s}^{-1}$$

After the string becomes slack (when the 7 kg mass is at rest on the table) the 3 kg mass moves as a free particle under gravity. If t is the time taken for this mass to come to rest, then using $v = u + at$

$$0 = 3.92 - gt \Rightarrow t = \frac{3.92}{g} = \frac{3.92}{9.8} = 0.4\,\text{s}$$

The 3 kg mass will come to rest instantaneously after 0.4 s.

Example 10

A light inextensible string is attached to the ceiling and passes under a moveable pulley of mass m kg and over a fixed pulley before having a mass of 4 kg tied to its free end. If the system is released from rest find (a) the value of m if the 4 kg mass has a downward acceleration of $\frac{1}{2}g\,\text{m}\,\text{s}^{-2}$ and (b) the acceleration of the 4 kg mass if $m = 4$. Show that the moveable pulley will descend if $m > 8$.

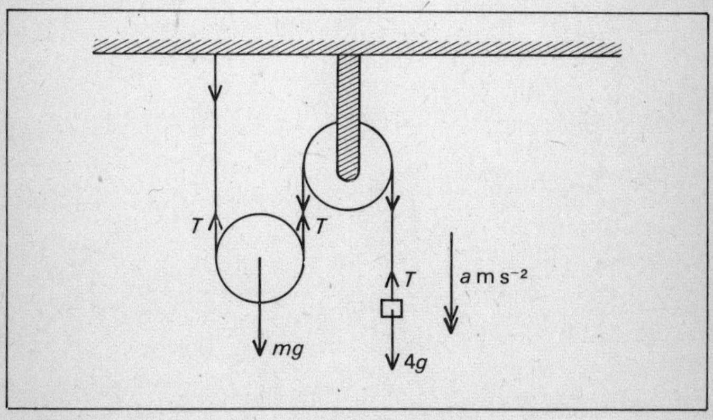

Figure 51

Let the tension in the string be T N and the acceleration of the 4 kg mass be $a \text{ m s}^{-2}$. Now if the 4 kg mass moves down a distance x then the pulley moves up a distance $\frac{1}{2}x$. Thus as the acceleration of the 4 kg mass is $a \text{ m s}^{-2}$, the acceleration of the pulley is $\frac{1}{2}a \text{ m s}^{-2}$. The system is shown in Figure 51.

The equations of motion are

on the 4 kg mass: $$4g - T = 4a \qquad (1)$$

on the m kg pulley: $$2T - mg = \tfrac{1}{2}ma \qquad (2)$$

Eliminating T from equations (1) and (2) gives

$$8g - mg = 8a + \tfrac{1}{2}ma$$

$$\Rightarrow \qquad a = \frac{2(8-m)g}{16+m} \qquad (3)$$

and $$m = \frac{16g - 16a}{2g + a} \qquad (4)$$

(a) If $a = \frac{1}{2}g$ then $m = \frac{16}{5}$ from (4).
Hence $m = \frac{16}{5}$ if the 4 kg mass has a downward acceleration of $\frac{1}{2}g$.

(b) If $m = 4$ then, using (3), $a = \frac{8}{20}g = \frac{2}{5}g$.
If the mass of the pulley is 4 kg then the acceleration of the 4 kg mass is $\frac{2}{5}g \text{ m s}^{-2}$.
Using (3) and assuming that $m > 0$, then $a < 0$ when $m > 8$ and hence the 4 kg mass will ascend and the pulley will descend if $m > 8$.

Chapter 18
Work, Power and Energy

The work done by a force is the product of the magnitude of the force and the distance moved in the direction of the force. This can be expressed as $\mathbf{F} \times \mathbf{r}$ where $\mathbf{r}$ is the displacement of the force $\mathbf{F}$. Work is measured in joules.

Power is defined as the rate at which work is done and is measured in watts.

The energy of a body is its capacity for doing work and is measured in joules. The two forms of energy which are required in this section are potential and kinetic energy and many questions also use the Principle of Work and the Conservation of mechanical energy. In these topics, motion takes place and thus the ideas used in Chapter 17 are often needed to solve the problems set.

Multiple choice questions (Type A) Select the correct answer

Example 1
The work done by the force $\mathbf{F} = 5\mathbf{i} - 4\mathbf{j} + 3\mathbf{k}$ when it moves its point of application through a displacement given by $\mathbf{r} = \mathbf{i} + 2\mathbf{j} + \mathbf{k}$ is

A 0 **B** $6\sqrt{3}$ **C** 16 **D** $10\sqrt{3}$ **E** none of these

The work done is given by $\mathbf{F} \cdot \mathbf{r}$.
$\therefore$ work done $= (5\mathbf{i} - 4\mathbf{j} + 3\mathbf{k}) \times (\mathbf{i} + 2\mathbf{j} + \mathbf{k}) = 5 - 8 + 3 = 0.$

ANSWER A

Remember that, in general, the answer is not $|\mathbf{F}| \times |\mathbf{r}|$, i.e. not $\sqrt{5^2 + 4^2 + 3^2} \times \sqrt{1^2 + 2^2 + 1^2} = \sqrt{50} \times \sqrt{6} = 10\sqrt{3}$.

This is only applicable when the force is in the direction of the displacement.

Example 2
A particle of mass m is allowed to slide a distance x down a rough plane inclined at 30° to the horizontal. If μ is the coefficient of friction then the total work done is

A $\frac{1}{2}mgx(2 - \mu\sqrt{3})$ **B** $\frac{\sqrt{3}}{2}mgx(1 + \mu)$ **C** $\frac{1}{2}mgx(1 - \mu\sqrt{3})$

D $\frac{1}{2}mgx[1 + \sqrt{3}(1 - \mu)]$ **E** $\frac{\sqrt{3}}{2}mgx(2 + \mu)$

The forces acting on the particle causing the displacement x are shown in Figure 52.

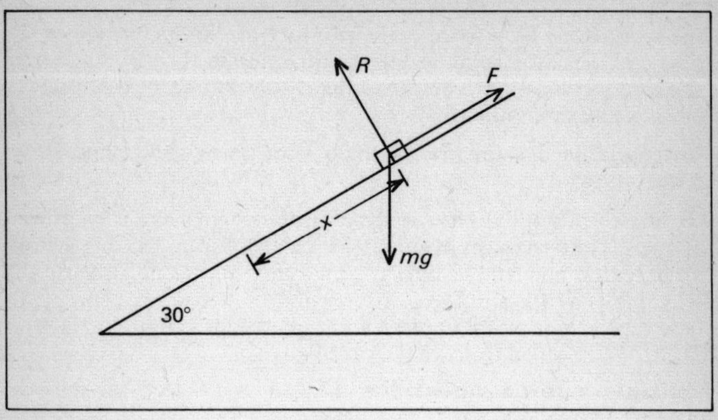

Figure 52

Since there is no motion perpendicular to the plane we have

$$R = mg \cos 30° = \tfrac{1}{2}\sqrt{3}mg$$

The work done (a) by R is $Rx \cos 90° = 0$; (b) by F is $Fx \cos 180°$ or $-Fx$; and (c) by mg is $mgx \sin 30° = \tfrac{1}{2}mgx$.
The total work done $= \tfrac{1}{2}mgx - Fx$
Now $F = \mu R = \tfrac{1}{2}\mu mg\sqrt{3}$ and hence,

$$\text{total work done} = \tfrac{1}{2}mgx - \tfrac{1}{2}\sqrt{3}\mu mgx = \tfrac{1}{2}mgx(1-\mu\sqrt{3})$$

ANSWER C

Example 3
A car of mass 1000 kg is moving along a straight horizontal road against constant resistances. If the pulling force of the engine is 450 N and it accelerates from 5 m s^{-1} to 10 m s^{-1} in 300 m then the resistances are

A 200 N **B** $241\tfrac{2}{3}$ N **C** 325 N **D** $441\tfrac{2}{3}$ N **E** none of these

The Principle of Work gives

$$\text{Work done} = \text{Final K.E.} - \text{Initial K.E.}$$

If the resistances are R N then the effective forward force on the car is $(450 - R)$ N.

Hence $(450-R)\times 300 = \frac{1}{2}\times 1000\times 10^2 - \frac{1}{2}\times 1000\times 5^2$

$\Rightarrow 300(450-R) = 500(100-25) \Rightarrow 450-R = 125$

giving $R = 325$ N. The resistances are 325 N. ANSWER C

Alternatively, this result can be obtained from Newton's second law after finding the acceleration.

i.e. using $v^2 = u^2 + 2as$ gives $10^2 = 5^2 + 2a(300) \Rightarrow a = \frac{1}{8}\,\text{m s}^{-2}$.

Using $F = ma$ we have, $450 - R = 1000 \times \frac{1}{8} \Rightarrow R = 325$ N.

Example 4

A particle of mass 4 kg is attached to one end of an elastic string of natural length 2 m and modulus $20g$ N. If the other end is fixed to a point A of the ceiling and the particle is projected vertically downwards with a speed of $\sqrt{2g}\,\text{m s}^{-1}$ from A, the displacement before the particle comes momentarily to rest is

A $1\frac{1}{5}$ m **B** 2 m **C** $3\frac{1}{5}$ m **D** 4 m **E** none of these

The particle falls freely under gravity during the first part of the motion when the string is slack. After the natural length has been reached, work is done in stretching the string and since there are no impulses in the motion the Principle of Work can be used.

If x is the displacement beyond the natural length when the particle comes to rest then the work done by gravity is $4g(x+2)$.

The work done by the tension in the string is $-\dfrac{\lambda x^2}{2a}$ where λ is the modulus and a is the natural length.

Work done by the string is $-\dfrac{20gx^2}{4} = -5gx^2$.

The kinetic energies at A and B are $\frac{1}{2}mu^2 = \frac{1}{2}(4)2g = 4g$ and zero respectively.

Since Work done = Final K.E. − Initial K.E.

$$4g(x+2) - 5gx^2 = 0 - 4g$$

$\Rightarrow 5x^2 - 4x - 12 = 0 \Rightarrow (5x+6)(x-2) = 0$

$\Rightarrow x = 2$ or $-1\frac{1}{5}$. Since $x > 0$, the solution is $x = 2$ and the displacement from A is 4 m. ANSWER D

General questions

Full examination questions often involve topics from other parts of the syllabus as well.

> **Example 5**
>
> A particle of mass m kg is placed at the highest point, A, of the surface of a fixed sphere of radius a and centre O. It is pushed so that it begins to move with a speed $\sqrt{2ga/5}$ down the surface of the sphere. If the particle leaves the surface of the sphere at P where angle $AOP = \theta$, show that $\cos\theta = \frac{4}{5}$. If $a = \frac{5}{8}$ find the horizontal displacement of the particle from the vertical through O when it reaches the horizontal plane on which the sphere is standing.

Let the reaction between the particle and the sphere be R and let v be the velocity at P. Now when the particle leaves the surface at P, $R = O$. Since it initially performs circular motion with centre O and radius a we have, the force required towards the centre $= \dfrac{mv^2}{a}$ and this is provided by $mg\cos\theta$.

Hence $$mg\cos\theta = \frac{mv^2}{a} \Rightarrow v^2 = ga\cos\theta \qquad (1)$$

Since the reaction between the particle and the surface is always perpendicular to the direction of motion it does no work. Thus mechanical energy is conserved.

i.e. Total energy at A = Total energy at P

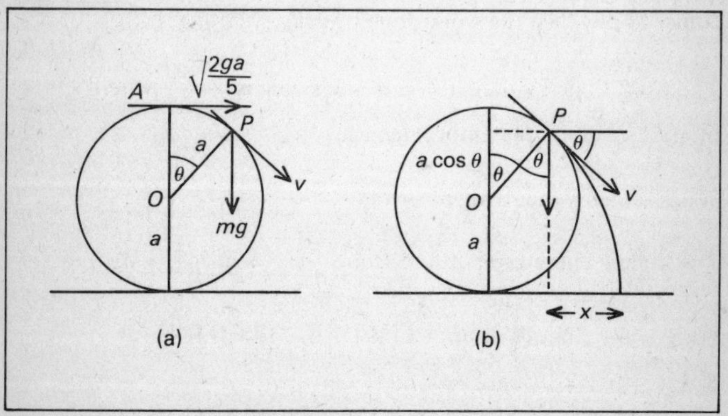

Figure 53

Regarding the zero level of potential energy as the level of O,

$$\tfrac{1}{2}m\left(\sqrt{\frac{2}{5}ga}\right)^2 + mga = \tfrac{1}{2}mv^2 + mga\cos\theta$$

$$\Rightarrow v^2 = \tfrac{2}{5}ga + 2ga(1-\cos\theta) \qquad (2)$$

Combining (1) and (2) gives

$$ga\cos\theta = \tfrac{2}{5}ga + 2ga(1-\cos\theta)$$

$$\cos\theta = \tfrac{2}{5} + 2 - 2\cos\theta \Rightarrow 3\cos\theta = \tfrac{12}{5} \Rightarrow \cos\theta = \tfrac{4}{5}$$

Hence the particle leaves the surface when $\cos\theta = \frac{4}{5}$.

After this the particle moves as a free projectile under gravity. Let t be the time of flight from P to the plane.
The components of the velocity at P are:

horizontally $\quad v\cos\theta = \sqrt{\dfrac{4ga}{5}} \times \tfrac{4}{5} = \tfrac{8}{5}\sqrt{\dfrac{ga}{5}}$

vertically $\quad v\sin\theta = \sqrt{\dfrac{4ga}{5}} \times \tfrac{3}{5} = \tfrac{6}{5}\sqrt{\dfrac{ga}{5}}$

If x is the horizontal displacement of the particle from P when it strikes the plane then since the velocity is constant,

$$x = \tfrac{8}{5}\sqrt{\frac{ga}{5}}\,t \Rightarrow x = \tfrac{8}{5}\sqrt{\frac{g}{8}}\,t \qquad \text{as } a = \tfrac{5}{8} \qquad (3)$$

Using $s = ut + \frac{1}{2}gt^2$ vertically

$$\Rightarrow \frac{9a}{5} = \tfrac{6}{5}\sqrt{\frac{ga}{5}}\,t + \tfrac{1}{2}gt^2 \qquad (4)$$

As $a = \frac{5}{8}$ this gives, using $t = \dfrac{5x}{8}\sqrt{\dfrac{8}{g}}$, (from (3))

$$\tfrac{9}{8} = \tfrac{6}{5}\sqrt{\frac{g}{8}} \times \frac{5x}{8}\sqrt{\frac{8}{g}} + \frac{g}{2} \times \frac{25x^2}{64} \times \frac{8}{g}$$

$$\Rightarrow \tfrac{9}{8} = \frac{6x}{8} + \frac{25x^2}{16} \Rightarrow 25x^2 + 12x - 18 = 0$$

Using the formula to solve gives $x = \dfrac{-12 \pm \sqrt{144+1800}}{50}$

$$\Rightarrow x = \frac{-12 \pm 44.09}{50} \Rightarrow x = 0.64 \text{ or } -1.12 \text{ to 2 decimal places.}$$

Since $x > 0$, choose the positive solution to give $x = 0.64$ m. The total displacement from the vertical through O is $x + a\sin\theta = 0.64 + (\frac{5}{8} \times \frac{3}{5}) = 1.0$ m to 1 decimal place.

Example 6
A train of total mass 150 tonnes is travelling along a level track at a constant speed of $96\,\text{km h}^{-1}$. If the resistances to motion are 50 N/tonne find the power developed by the engine.
If the engine continues to work at this rate find the acceleration of the train on an incline of 1 in 100 ($\sin^{-1}\frac{1}{100}$) when the speed is $18\,\text{km h}^{-1}$, assuming that the resistances remain constant. Take the value of g as $10\,\text{m s}^{-2}$.

The total resistances are $50 \times 150 = 7500$ N.
At maximum speed the tractive effort is equal to the resistances.
If P N is the tractive effort then, when the speed is $v\,\text{m s}^{-1}$, the work done per second (power) is Pv watts.

Hence Power $= 7500 \times 96 \times \frac{5}{18} = 2500 \times 80 = 200$ kW

Since the power remains constant, the tractive force produced by the engine at a speed of $18\,\text{km h}^{-1}$ is $\dfrac{200\,000}{(18 \times \frac{5}{18})} = 40\,000$ N.

The effective forward force on the train is

Tractive force − Resistances − Weight component
$= 40\,000 - 7500 - (150\,000g \times \frac{1}{100}) = 17\,500$ N.

Using Newton's second law

$$17\,500 = 150\,000a \Rightarrow a = \frac{175}{1500} = 0.12\,\text{m s}^{-2} \text{ to 2 decimal places.}$$

The acceleration on the incline is $0.12\,\text{m s}^{-2}$ to 2 decimal places.

Example 7
A particle of mass 2 kg is attached to the mid-point of an elastic string AB of natural length 4 m and modulus $12g$ N. If A and B are fixed at the same level 4 m apart and the particle is released from rest at this level from the mid-point of AB find the speed of the particle after it has fallen $1\frac{1}{2}$ m.

Initially the string is just taut and when released from the mid-point of AB the particle will fall vertically.

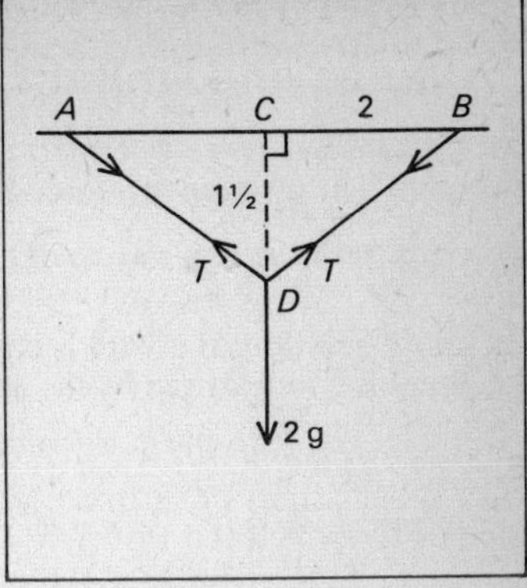

Figure 54

From Figure 54, by using the Theorem of Pythagoras in $\triangle BCD$,

$$BD^2 = 2^2 + (1\tfrac{1}{2})^2 \Rightarrow BD = \tfrac{5}{2}$$

Now the work done in stretching an elastic string $= \frac{1}{2}\lambda x^2/l$ where x is the extension and l is the natural length. Therefore,

$$\text{work done} = \frac{-12g(1)^2}{2 \times 4}$$

$\Rightarrow -\frac{12}{8}g = -\frac{3}{2}g$ J.

Now the work done by gravity is

$2g(1\frac{1}{2}) = 3g$ J.

By the Principle of Work,

$$\text{Work done} = \text{Final K.E.} - \text{Initial K.E.}$$

$3g - \frac{3}{2}g = \frac{1}{2} \times 2 \times v^2 \Rightarrow v^2 = \frac{3}{2}g \Rightarrow$ the velocity is $\sqrt{\frac{3}{2}g}$ m s^{-1}.

Chapter 19
Impulse and Impact

A constant force is said to exert an impulse when it acts for a time t seconds. The impulse of a force $\mathbf{F}$ is $\mathbf{F}t$ and is measured in Newton seconds. If the force is variable then the impulse is given by $\int_0^t \mathbf{F}\,dt$.

In many questions the force is large and the time interval short and the impulse is measured by the effect it produces, namely

$$\text{Impulse} = \text{Change in momentum}$$

The other aspect of this section is concerned with the impact of bodies for which the Principle of Conservation of Momentum is used, i.e., if the vector sum of the external forces acting on a system is zero then the total momentum is constant.

Multiple choice questions (Type A) Select the correct answer

Example 1
A particle is acted upon by a force $\mathbf{F}$ given by $(2t-1)\mathbf{i}+3t^2\mathbf{j}$ for a time interval from $t = 3$ s to $t = 5$ s. The magnitude of the impulse of the force in Newton seconds is

A 12 **B** $\sqrt{153}$ **C** $26\sqrt{5}$ **D** $70\sqrt{2}$ **E** none of these

Since the force is variable, $\text{Impulse} = \int_3^5 \mathbf{F}\,dt$

$$\therefore \quad \text{Impulse} = \int_3^5 (2t-1)\mathbf{i}+3t^2\mathbf{j}\,dt = [(t^2-t)\mathbf{i}+t^3\mathbf{j}]_3^5$$

$$= (20\mathbf{i}+125\mathbf{j})-(6\mathbf{i}+27\mathbf{j}) = 14\mathbf{i}+98\mathbf{j}$$

The magnitude of the impulse =

$$14\sqrt{1^2+7^2} = 14\times\sqrt{50} = 70\sqrt{2}.$$

ANSWER D

Example 2
A sphere A, of mass m kg, moving with a speed of $2\,\text{m}\,\text{s}^{-1}$ collides directly with a second sphere B, of mass 2 kg moving in the opposite direction with a speed of $4\,\text{m}\,\text{s}^{-1}$. If sphere A is brought to rest by the impact and the coefficient of restitution is $\frac{1}{2}$, the value of m is

A 1 **B** 3 **C** 5 **D** 7 **E** none of these

Let the final speed of B be v and assume that the original direction of motion of A is positive.

By Conservation of momentum: $2m - 8 = 2v$ (1)

By Newton's Experimental law: $v - 0 = -\frac{1}{2}(-4-2)$ (2)

From (2) $v = 3\,\text{m s}^{-1}$

Hence using (1) $2m - 8 = 6 \Rightarrow m = 7$

ANSWER D

Remember that momentum and velocity are vectors and the correct sign must be applied in Equations (1) and (2).

Example 3
A ball strikes a wall while its direction of motion makes an angle of 60° with the wall. If it rebounds at an angle of 30° to the wall, the coefficient of restitution is

A 0 **B** $\frac{1}{3}$ **C** $\frac{1}{2}$ **D** $\frac{2}{3}$ **E** 1

Let the velocity of the ball before striking the wall be $v\,\text{m s}^{-1}$ at 60° to the wall. The components of this velocity parallel and perpendicular to the wall are $v\cos 60°$ and $v\sin 60°$ respectively.

Only the component perpendicular to the wall will be affected by the impact. This will become $ev\sin 60°$ away from the wall, where e is the coefficient of restitution.

Hence, after the impact, the ball has velocity components of $v\cos 60°$ and $ev\sin 60°$ parallel and perpendicular to the wall respectively. Thus if the direction of motion is at 30° to the wall

$$\tan 30° = \frac{ev\sin 60°}{v\cos 60°} = e\tan 60° \Rightarrow e = \tfrac{1}{3}$$

ANSWER B

Example 4
Two particles A and B of masses 2 kg and 4 kg respectively lie on a smooth horizontal table connected by a light, taut, inextensible string. If B is given an impulse of $12\sqrt{2}$ N s in a direction making an angle of 45° with AB then the direction of motion of B immediately after the impulse has been applied makes an angle θ with AB where $\tan\theta$ equals

A 0 **B** $\frac{2}{3}$ **C** 1 **D** $\frac{3}{2}$ **E** none of these

Let the components of the velocity of B after the impulse be u and v along and perpendicular to AB as shown in Figure 55. As the string is initially taut, A will also begin to move with a velocity u along AB.

The components of the impulse along and perpendicular to AB are $12\sqrt{2}\cos 45°$ and $12\sqrt{2}\sin 45°$.

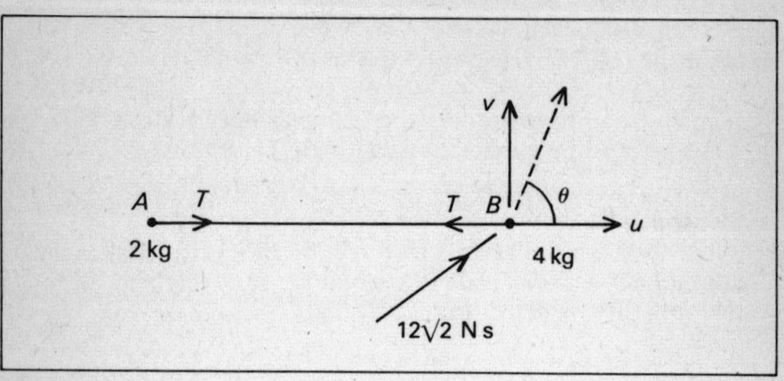

Figure 55

Since Impulse = Change in momentum in that direction

Impulse along AB:

$$12\sqrt{2}\cos 45° = 4u+2u \qquad (1)$$

Impulse normal to AB:

$$12\sqrt{2}\sin 45° = 4v \qquad (2)$$

From (1) and (2) $\quad 6u = 12 \Rightarrow u = 2\,\text{m s}^{-1}$

and $\quad 4v = 12 \Rightarrow v = 3\,\text{m s}^{-1}$

Thus the direction of motion is given by $\tan\theta = \dfrac{v}{u} = \frac{3}{2}$.

ANSWER D

General questions

Full examination questions often combine several different aspects of the syllabus as illustrated by the following examples.

Example 5

A small block of wood of mass 4 kg hangs at rest on the end of a light, inextensible string of length 1 m with its other end attached to a fixed point, O, of the ceiling. A pellet of mass 20 g is fired horizontally with a speed of $804\,\text{m s}^{-1}$ at the block and becomes embedded in it. Find the angle that the string makes with the downward vertical when the block comes to instantaneous rest. Take the value of g as $9.8\,\text{m s}^{-2}$.

Let the speed of the combined mass after the pellet enters the block be $V\,\mathrm{m\,s^{-1}}$. Since, at the moment of impact, momentum is conserved

$$0.02 \times 804 = 4.02V \Rightarrow 1.608 = 4.02V \Rightarrow V = 4\,\mathrm{m\,s^{-1}}$$

After the impact, the combined mass moves in a circle, centre O, and mechanical energy is conserved. If θ is the angle that the string makes with the downward vertical when the block comes to rest

$$\tfrac{1}{2}mV^2 = mg(1-\cos\theta) \Rightarrow \tfrac{1}{2} \times 16 = 9.8(1-\cos\theta)$$

$$1-\cos\theta = 0.8163 \Rightarrow \cos\theta = 0.1837 \Rightarrow \theta = 79°\,25'$$

The string makes an angle of 79° 25′ with the vertical.

Example 6

A particle A, of mass 3 kg, rests on a smooth plane inclined at 30° to the horizontal and is connected by a light, inextensible string passing over a smooth pulley at the top of the incline to a particle B, of mass 2 kg, hanging vertically. The system is released from rest. If, after falling 1 m, B hits an inelastic table find the distance A moves before coming to instantaneous rest and the time that elapses before the string is again taut. Find also the impulsive tension in the string when B is set in motion again.

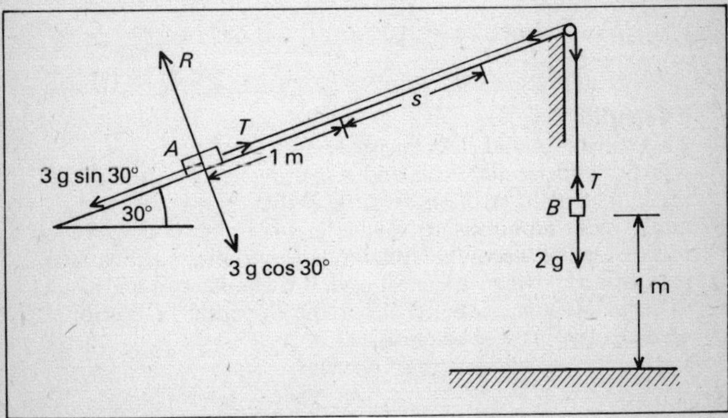

Figure 56

Let the acceleration of the system be a and the tension in the string be T. Using Newton's second law on 2 kg mass: $2g - T = 2a$

on 3 kg mass: $T - 3g\sin 30° = 3a$

Adding gives $\frac{1}{2}g = 5a \Rightarrow a = \frac{1}{10}g \text{ m s}^{-2}$

Now if the velocity of the system after the masses have moved 1 m is v then using $v^2 = u^2 + 2as$

$$v^2 = 0 + 2 \times 1 \times \frac{g}{10} \Rightarrow v = \sqrt{\frac{g}{5}} = 1.4 \text{ m s}^{-1}$$

At this point, B strikes the table without rebounding and the string becomes slack. Thus A moves with an acceleration of $-g\sin 30°$.

Using $v^2 = u^2 + 2as \Rightarrow 0 = \frac{g}{5} - 2\left(\frac{g}{2}\right)s \Rightarrow s = \frac{1}{5}$ m.

Using $v = u + at \Rightarrow 0 = 1.4 - \frac{1}{2}gt \Rightarrow t = \dfrac{2.8}{9.8} = \frac{2}{7}$ s.

So it is clear that A moves $\frac{1}{5}$ m in $\frac{2}{7}$ s after the string is initially slack.

Let V be the speed of both masses after B is set in motion again. On the whole system, momentum is conserved

$$5V = 3\sqrt{\frac{g}{5}} \Rightarrow V = \tfrac{3}{5} \times 1.4 = 0.84\, m\, s^{-1}$$

Thus, considering B, Impulse = Change in momentum

and Impulse $= 2[0.84 - 0] = 1.68$ N s

Example 7

Two spheres A and B, of masses m and $2m$ respectively, lie on a smooth horizontal table and A is projected towards B with speed u so as to strike it directly. B then hits a wall at right angles and rebounds to hit A again. If the coefficient of restitution between the spheres is $\frac{1}{4}$ find the coefficient of restitution between the wall and B if the magnitude of the velocity of A is unchanged but the direction of motion is reversed by this second collision.

Consider the first impact (see Figure 57). Let v_1 and v_2 be the speeds of A and B respectively after the impact.

By conservation of momentum

$$mv_1 + 2mv_2 = mu \Rightarrow v_1 + 2v_2 = u \qquad (1)$$

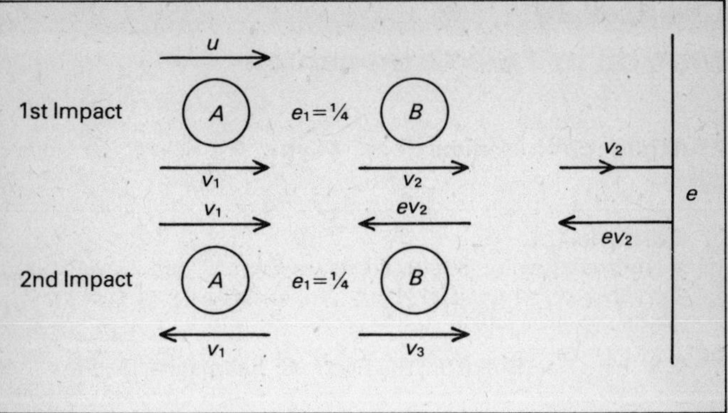

Figure 57

By Newton's Experimental law

$$v_2 - v_1 = -e_1(0-u) \Rightarrow v_2 - v_1 = \tfrac{1}{4}u \qquad (2)$$

Solving (1) and (2) by adding gives $v_2 = \frac{5}{12}u$.
Using (2) gives $v_1 = \frac{1}{6}u$.

Sphere B now moves with constant speed until it strikes the wall from which it rebounds with speed $\frac{5}{12}eu$.

Consider the second impact (see Figure 57). The final speed of A will be v_1 as shown. Let the final speed of B be v_3.

By Conservation of Momentum

$$2mv_3 - mv_1 = mv_1 - 2m(ev_2)$$

$$\Rightarrow 2v_3 = \frac{u}{6} - \frac{5eu}{6} + \frac{u}{6} \Rightarrow v_3 = \frac{u}{6} - \frac{5eu}{12} \qquad (3)$$

By Newton's Experimental law

$$v_3 + v_1 = -e_1(-ev_2 - v_1)$$

$$v_3 = -\tfrac{1}{4}\left(-\frac{5eu}{12} - \frac{u}{6}\right) - \frac{u}{6} \Rightarrow v_3 = \frac{5eu}{48} - \frac{u}{8} \qquad (4)$$

Equating (3) and (4)

$$\frac{5eu}{48} - \frac{u}{8} = \frac{u}{6} - \frac{5eu}{12}$$

$$\Rightarrow 5e - 6 = 8 - 20e \Rightarrow 25e = 14 \Rightarrow e = \tfrac{14}{25}$$

The coefficient of restitution between the wall and B is $\frac{14}{25}$.

Chapter 20
Motion in Two Dimensions

Multiple choice questions (Type A) Select the correct answer

Example 1
A wind is blowing from the East at $3\,\mathrm{m\,s^{-1}}$ and a cyclist is travelling due North at $4\,\mathrm{m\,s^{-1}}$. The velocity of the wind relative to the cyclist

A is $1\,\mathrm{m\,s^{-1}}$ **B** is from the East **C** has magnitude $5\,\mathrm{m\,s^{-1}}$

D $-1\,\mathrm{m\,s^{-1}}$ **E** is from the West

The velocity of the wind relative to the cyclist is

$$_{\mathrm{w}}\mathbf{v}_{\mathrm{c}} = \mathbf{v}(\text{wind}) - \mathbf{v}(\text{cyclist}) = \mathbf{v}_{\mathrm{w}} - \mathbf{v}_{\mathrm{c}}$$

Remembering that these velocities are vectors, and using $\mathbf{i}$(East) and $\mathbf{j}$(North), $\mathbf{v}_{\mathrm{w}} = -4\mathbf{i}$, $\mathbf{v}_{\mathrm{c}} = 3\mathbf{j} \Rightarrow {}_{\mathrm{w}}\mathbf{v}_{\mathrm{c}} = -4\mathbf{i} - 3\mathbf{j}$ which has magnitude $5\,\mathrm{m\,s^{-1}}$. ANSWER C

Example 2
A particle moving in a circle of radius 3 metres has angular velocity 3 revolutions per minute. Its velocity along the tangent is

A $1\,\mathrm{m\,s^{-1}}$ **B** $1\,\mathrm{m\,min^{-1}}$ **C** $9\,\mathrm{m\,s^{-1}}$ **D** $3\,\mathrm{m\,min^{-1}}$
E $\frac{3}{10}\pi\,\mathrm{m\,s^{-1}}$

$v = r\omega$ and if r is in metres and ω in radians per second then v will be in $\mathrm{m\,s^{-1}}$.
$\omega = 3\,\text{rev/min} = 3 \times 2\pi\,\text{radians/min} = \frac{6}{60}\pi\,\mathrm{rad\,s^{-1}} \doteq \frac{1}{10}\pi^{c}\,\mathrm{s^{-1}}$
$v = r\omega = 3 \times \frac{1}{10}\pi = \frac{3}{10}\pi\,\mathrm{m\,s^{-1}}$. ANSWER E

Example 3
The particle of Example 2 has acceleration towards the centre of

A $27\,\mathrm{m\,s^{-2}}$ **B** $9\,\mathrm{m\,s^{-2}}$ **C** $3\,\mathrm{m\,s^{-2}}$ **D** $\frac{3}{100}\pi^2\,\mathrm{m\,s^{-2}}$
E $\frac{9}{10}\pi\,\mathrm{m\,s^{-2}}$

$$a = r\omega^2 = 3 \times \frac{\pi^2}{100} = \frac{3\pi^2}{100}\,\mathrm{m\,s^{-2}}$$ ANSWER D

Multiple choice questions (Type B) Answer according to the table

A	**B**	**C**	**D**	**E**
1, 2, 3 correct	1, 3 only	2, 3 only	2 only	3 only

Example 4
For a projectile with starting speed u at an angle θ above the horizontal

1 Greatest height $= u^2\sin^2\theta/2g$ **2** Range $R = u\sin^2 2\theta/g$
3 Time of flight $T = 2u\sin\theta/g$

The range is the distance the projectile travels on a horizontal plane and the time taken for this is T.
With the usual notation: x for horizontal and y for vertical displacement from the starting point at time t,

$$x = u\cos\theta \times t,\ y = u\sin\theta \times t - \tfrac{1}{2}gt^2,\ \dot{y} = \frac{dy}{dt} = u\sin\theta - gt$$

Time to reach greatest height is given by $\dot{y} = 0 \Rightarrow t = u\sin\theta/g$
The greatest height is the value of y at this time

$$Y = u\sin\theta \cdot \frac{u\sin\theta}{g} - \tfrac{1}{2}g \cdot \frac{u^2\sin^2\theta}{g^2}$$

$$= u^2\sin^2\theta/g - u^2\sin^2\theta/2g = \frac{u^2\sin^2\theta}{2g}$$

Time of flight $T = 2t = 2u\sin\theta/g$. As a check $y = 0$ gives the same value.
Horizontal range R is given by the x value at this time

$$R = u\cos\theta \times t = u\cos\theta \cdot \frac{2u\sin\theta}{g} = \frac{u^2\sin 2\theta}{g}$$

Statements 1, 2 and 3 are correct. ANSWER A

Example 5
The coordinates of a particle P at time t are $x = 3\cos 2t$ and $y = 3\sin 2t$.

1 P moves in a circle of radius 3
2 The angular velocity of P is 2 rads/s
3 The acceleration of P has magnitude $12\,\mathrm{m\,s^{-2}}$

$x^2+y^2=9\cos^2 2t+9\sin^2 2t=9$ which is a circle of radius 3.
P describes this circle while $2t$ changes by 2π, i.e., in π seconds.

Its angular velocity ω is $\dfrac{2\pi}{\pi}=2$ radians/second.

The acceleration (towards the centre) is $r\omega^2=3\times 2^2=12\,\text{m s}^{-2}$
Statements 1, 2 and 3 are all correct. ANSWER A

General questions

> **Example 6**
> A car negotiates a banked circular curve of radius 100 m and angle of inclination 30°. The coefficient of friction between the tyres and the road surface is $\frac{2}{3}$. Find the greatest speed before the car tends to move up the curve.

For the car to slip up the slope, friction acts down the slope (Figure 58).

$\uparrow$ forces balance $\Rightarrow$
$R\cos 30° = F\sin 30° + mg$
$F=\mu R \Rightarrow R\cdot\frac{1}{2}\sqrt{3}-\frac{2}{3}\cdot R\cdot\frac{1}{2}=mg$

$$\Rightarrow R=\frac{6mg}{3\sqrt{3}-2}$$

$\leftarrow F=ma \Rightarrow F\cos 30° + R\sin 30°$

$$=\frac{mv^2}{100}$$

$$\frac{mv^2}{100}=\tfrac{2}{3}R\cdot\tfrac{1}{2}\sqrt{3}+R\cdot\tfrac{1}{2}$$

$$=\frac{6mg}{(3\sqrt{3}-2)}\left(\tfrac{1}{3}\sqrt{3}+\tfrac{1}{2}\right)$$

$$v^2=\frac{100g(2\sqrt{3}+3)}{3\sqrt{3}-2}=\frac{981\times 6.464}{3.196}$$

$v^2=1984 \Rightarrow v=44.5\text{ ms}=100\text{ mph}$

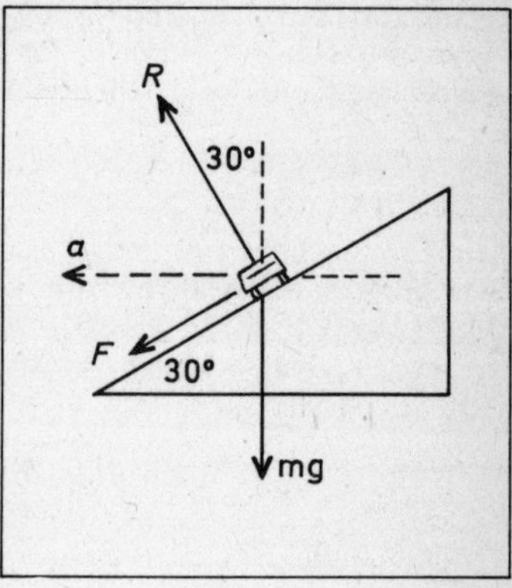

Figure 58

> **Example 7**
> When fielding on the boundary (at cricket) how fast must I throw the ball for it to be able to reach the wicketkeeper (60 m away) without bouncing?

Assuming that the wicketkeeper catches the ball at the same height from which I throw it, the maximum range for a projectile is

$$R = \frac{v^2}{g} \text{ so } 60g = v^2 \Rightarrow v = 24.25\,\text{m s}^{-1}\,(= 87.3\,\text{km/hr}) = 54.6\,\text{mph}.$$

The angle of projection for maximum range is $\theta = 45°$.

The time taken is $T = \dfrac{2v \sin\theta}{g} = 3.5$ seconds.

To throw the ball in quicker, the angle of projection is lower and the throwing speed is greater, e.g.

$$\theta = 30°,\ R = 60\,\text{m} = \frac{v^2 \sin 2\theta}{g} \Rightarrow v^2 = \frac{60g}{\sin 60°} = 679$$

$$\Rightarrow v = 26.1\,\text{m s}^{-1}\,(= 58.6\,\text{mph}).$$

> **Example 8**
> How fast is a hammer thrower turning when he releases the 'hammer' on its flight of 200 m?

Assuming maximum range is given by $R = v^2/g \Rightarrow v^2 = 200g \Rightarrow v = 44.27\,\text{m s}^{-1}$. Assuming that the hammer rotates at a distance of 1 metre from the axis of rotation $\omega = \dfrac{v}{r} = 44.27^c\,\text{s}^{-1} = 7$ revs/s, which seems amazingly fast.

However, some of the energy given to the 'hammer' is due to the fact that the 'hammer' is accelerating and not all its velocity comes from pure rotation.

> **Example 9**
> In firing at a target 60 m away, how high above the horizontal must an archer aim his arrow to hit the 'bull' which is at the same height above the ground as that at which the arrow is released?

Take $v = 100$ mph ($= 44.\dot{4}\,\text{m s}^{-1}$).

$$R = \frac{v^2 \sin 2\alpha}{g} \Rightarrow \frac{60g}{v^2} = \sin 2\alpha = 0.2976 \Rightarrow 2\alpha = 17.3° \Rightarrow \alpha = 8.66°.$$

The faster the arrow the lower the angle of projection.

Example 10
Water is projected from a hose at a speed of $20\,\text{m s}^{-1}$. At what angle must the hose be aligned to clear a wall 4 m high, 30 m from the hose? (Assume the hose is 1 m off the ground).

Take $g = 10\,\text{m s}^{-2}$

$$x = 20\cos\alpha t \Rightarrow 30 = 20\cos\alpha \cdot t = \frac{3}{2\cos\alpha}$$

$$y = 3 = 20\sin\alpha \cdot t - \tfrac{1}{2}gt^2$$

$$3 = \frac{20\sin\alpha 3}{2\cos\alpha} - \frac{10}{2}, \frac{9}{4\cos^2\alpha}, \quad 3 = 30\tan\alpha - \tfrac{45}{4}\sec^2\alpha \Rightarrow 1$$

$$= 10\tan\alpha - \tfrac{15}{4}(1+\tan^2\alpha)$$

$$4 + 15 = 40\tan\alpha - 15\tan^2\alpha, \; 15t^2 - 40t + 19 = 0,$$

$$t = \frac{40 \pm \sqrt{1600 - 4 \times 15 \times 19}}{30}$$

$$t = \frac{40 \pm 2\sqrt{400 - 285}}{30} = \frac{40 \pm 21.44}{30} = 2.048 \;\text{ or }\; 0.6187$$

$\therefore \; \alpha = 64°$ or $31.7°$

Example 11
An aircraft A on a course of 043° flying at $700\,\text{km hr}^{-1}$ is 50 km due West of plane B flying at $800\,\text{km hr}^{-1}$ on a course 310°. Find their closest approach.

$$\mathbf{v}_A = 700\sin 43°\mathbf{i} + 700\cos 43°\mathbf{j}$$
$$= 477.4\mathbf{i} + 511.9\mathbf{j}$$

$$\mathbf{v}_B = -800\cos 40°\mathbf{i} + 800\sin 40°\mathbf{j}$$
$$= -612.8\mathbf{i} + 514.2\mathbf{j}$$

The velocity of B relative to A in the direction of BK (Figure 59) is

$${}_B\mathbf{v}_A = \mathbf{v}_B - \mathbf{v}_A = -1090\mathbf{i} + 2.3\mathbf{j}$$

$$\tan\alpha = \frac{2.3}{1090} = 0.002\,11$$
$$\Rightarrow \alpha = 0.121°$$

Closest approach $AK = AB\sin\alpha$.
$AK = 50 \times \sin 0.121° = 0.105\,\text{km}$.
The time to reach this point =

$$\frac{Bk}{|{}_B\mathbf{v}_A|} \simeq \frac{50}{1090} = 2.75 \text{ minutes.}$$

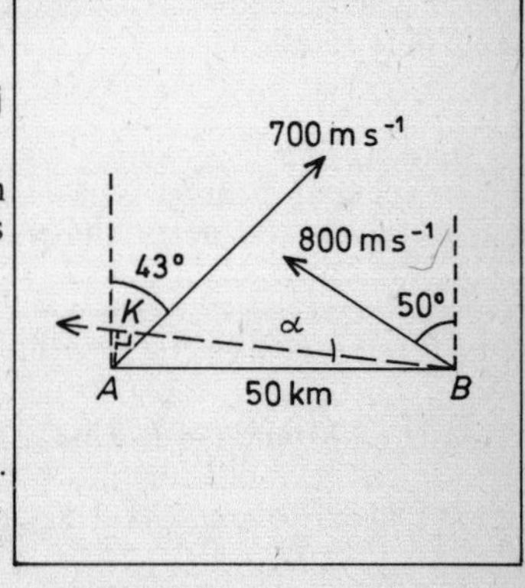

Figure 59

Example 12
On a journey from A to B (due North), to a cyclist travelling at $16\,\text{km hr}^{-1}$ the wind appears to come from the NE. When he returns from B to A at $20\,\text{km hr}^{-1}$ the wind appears to come from the East. Find the true velocity of the wind.

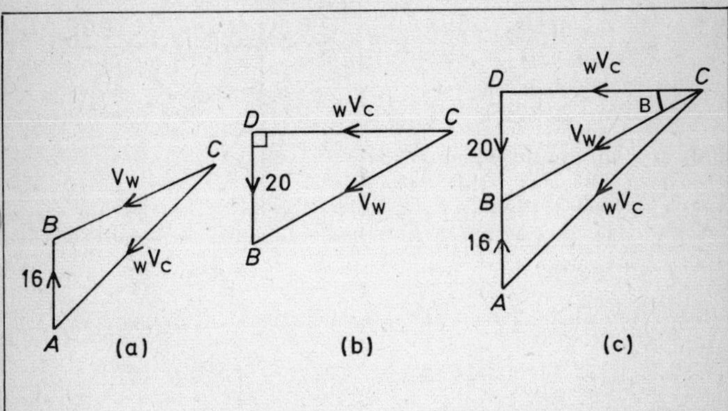

Figure 60

Using ${}_{\mathrm{w}}\mathbf{V}_{\mathrm{c}} = \mathbf{V}_{\mathrm{w}} - \mathbf{V}_{\mathrm{c}}$, Figures 60(a) and (b) show the velocities when the cyclist is travelling north and south respectively, and (c) puts these together.

In $\triangle ADC: DC = AD \tan 45° = 36\,\text{km hr}^{-1}$
In $\triangle BDC: BC^2 = 20^2 + 36^2 = 400 + 1296 = 1696$

$$BC = 41.2\,\text{km hr}^{-1}$$

$$\tan\beta = \tfrac{20}{36} = 0.5555 \Rightarrow \beta = 29°$$

The wind blows at $41.2\,\text{km hr}^{-1}$ from a direction N 61° E.

Example 13
A ship S is travelling at 10 knots due North. At 9 a.m. another ship T is sighted 5 nautical miles due East and at 10 a.m. T is 5 n.m. due South of S. Find the velocity of T, the shortest distance between S and T and the time at which this occurs.

Method 1

From Figure 61, T moves from $T_0(5,0)$ to $T_1(0,5)$ in 1 hour, so $\mathbf{v}_T = -5\mathbf{i}+5\mathbf{j} = 5\sqrt{2}$ knots NE. The position of S at time t is $10t\mathbf{j}$ and that of T, $5\mathbf{i}+t(-5\mathbf{i}+5\mathbf{j})$. The distance ST at time t is given by

$$\begin{aligned}\mathbf{ST} &= 5\mathbf{i}+t(-5\mathbf{i}+5\mathbf{j})-10t\mathbf{j}\\ &= (5-5t)\mathbf{i}-5t\mathbf{j}\\ |\mathbf{ST}| &= 25(1-t)^2+25t^2\\ &= 25(1-2t+2t^2)\end{aligned}$$

This is a minimum when $-2+4t=0 \Rightarrow t=\frac{1}{2}$. At this time (9.30 a.m.) S is at (0,5) and T is at $(2\frac{1}{2},2\frac{1}{2})$. The closest distance is $2\frac{1}{2}\sqrt{2} = 3.54$ n.m.

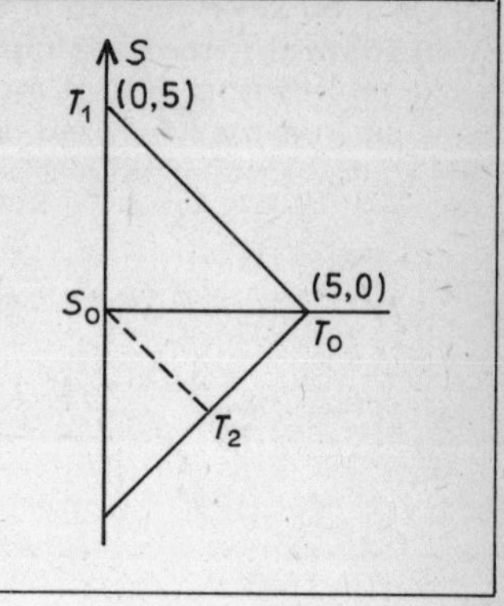

Figure 61

Method 2

The velocity of T relative to S is $\mathbf{v}_T-\mathbf{v}_S = -5\mathbf{i}+5\mathbf{j}-10\mathbf{j}$
$= -5\mathbf{i}-5\mathbf{j}$

Reducing S to rest (regarding S as fixed) T appears to move in the direction $-\mathbf{i}-\mathbf{j}$ (SW) from T_0 with velocity $-5\mathbf{i}-5\mathbf{j}$. The closest approach of T to S occurs when T is at T_2, i.e., when $t=\frac{1}{2}$ (when ST is perpendicular to T_0T_2).

This occurs at 9.30 a.m. and the closest approach is $S_0T_2 = 2\frac{1}{2}\sqrt{2} = 3.54$ n.m.

In general problems Method 2 is usually more straightforward whereas when the problem is set out using easy directions and Cartesian coordinates Method 1 may appear the easier.

Example 14

On a Big Wheel at a fairground how fast does the wheel (radius 10 m) turn to produce a feeling of 'weightlessness' at the top?

At the bottom, if the reaction from the chair is R then $R-mg = \dfrac{mv^2}{r}$. So at the top for 'weightlessness' $R=0$ and the person's weight produces the mass × acceleration.

$$mg = \frac{mv^2}{r} \Rightarrow v^2 = gr = 10g = 98.1 \Rightarrow v \simeq 10\,\mathrm{m\,s^{-1}}$$

The angular velocity $\frac{v}{r} = 1^c\,\mathrm{s}^{-1} \Rightarrow 1$ revolution in 6.25 seconds. $v = 10\,\mathrm{m\,s^{-1}} = 22.5$ mph is quite fast in the open air!

Example 15
A particle is projected at an angle of 60° above the horizontal with speed $20\,\mathrm{m\,s^{-1}}$. Find the greatest height it reaches and its range (a) on a horizontal plane (b) up a plane inclined at 30° above the horizontal and (c) down a plane inclined at 30° below the horizontal.

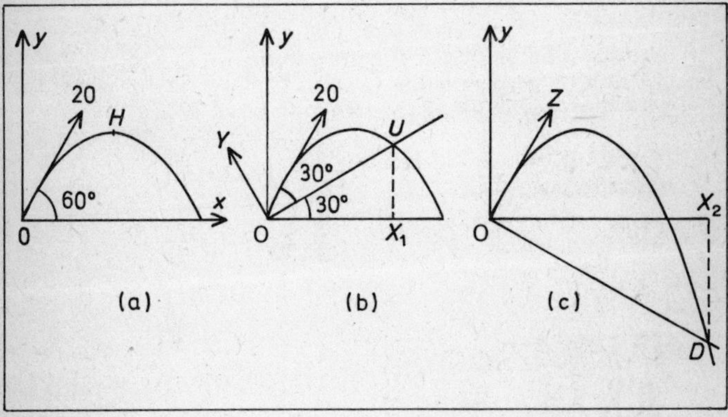

Figure 62

The greatest height reached is the same in all three cases. This problem is best solved by considering the horizontal and vertical motions separately. Horizontally there is no acceleration.

$$v_x = \dot{x} = 20\cos 60° = 10\,\mathrm{m\,s^{-1}} \qquad x = 20\cos 60° t = 10t$$

Vertically the acceleration is constant, $a = g = 9.8\,\mathrm{m\,s^{-1}}$.

$$v = u + at \Rightarrow \dot{y} = 20\sin 60° - gt = 10\sqrt{3} - gt$$
$$s = ut + \tfrac{1}{2}at^2 \Rightarrow y = 10\sqrt{3}t - \tfrac{1}{2}gt^2$$

At H, the highest point, $\dot{y} = 0 \Rightarrow gt = 10\sqrt{3} \Rightarrow t = 10\sqrt{3}/g$. The greatest height $y = \dfrac{10\sqrt{3}\cdot 10\sqrt{3}}{g} - \tfrac{1}{2}g\cdot\left(\dfrac{10\sqrt{3}}{g}\right)^2 = \dfrac{\tfrac{1}{2}\times 100\times 3}{g}$

$$= 15.3\text{ m.}$$

(a) The range R is the value of x when $t=T$, given by $y=0$, $T=\frac{20\sqrt{3}}{g}$.

$$R=10T=\frac{10\times 20\sqrt{3}}{g}=\frac{200\sqrt{3}}{g}=35.4\,\text{m}$$

(b) Range up the inclined plane (Figure 62(b)).
To find the time to reach U consider the motion perpendicular to OU, i.e., along OY where the acceleration is $-g\cos 30°$ so $s=ut+\frac{1}{2}gt^2 \Rightarrow Y=20\sin 30°t-\frac{1}{2}g\cos 30°t^2 \Rightarrow t=0$ (at O) or $t=\frac{40\tan 30°}{g}$ at U. $OX_1=10t=\frac{400\tan 30°}{g}$ and $OU=\frac{OX_1}{\cos 30°}=\frac{400}{g\sqrt{3}}\times\frac{2}{\sqrt{3}}=27.3\,\text{m}$.

(c) Range down the slope OD (Figure 62(c)).
Consider motion perpendicular to OD, i.e., along OZ for which the acceleration is $-g\cos 30°$ and the starting speed $20\,\text{m}\,\text{s}^{-1}$.

$$Z=20t-\tfrac{1}{2}g\cos 30°t^2=0 \text{ at } D \text{ where } t=\frac{40}{g\cos 30°}=\frac{80}{g\sqrt{3}}$$

$$OX_2=10t=\frac{800}{g\sqrt{3}}=47.1\,\text{m}\Rightarrow OD=\frac{OX_2}{\cos 30°}=47.1/0.866=54.4\,\text{m}$$